Dynamical Systems-Based Soil Mechanics

Dynamical Systems-Based Soil Mechanics

Paul G. Joseph

CRC Press
Taylor & Francis Group
Boca Raton   London   New York   Leiden

CRC Press is an imprint of the
Taylor & Francis Group, an **informa** business

A BALKEMA BOOK

Cover photographer
Sand from Hokkaido island, Japan by Catalin Stefan | World Atlas of Sands |
www.sand-atlas.com

CRC Press/Balkema is an imprint of the Taylor & Francis Group, an informa business

© 2017 Taylor & Francis Group, London, UK

Typeset by MPS Limited, Chennai, India
Printed and bound in the United States of America by Edwards Brothers Malloy
on sustainably sourced paper

Library of Congress Cataloging-in-Publication Data

Applied for

Published by: CRC Press/Balkema
                P.O. Box 11320, 2301 EH Leiden, The Netherlands
                e-mail: Pub.NL@taylorandfrancis.com
                www.crcpress.com – www.taylorandfrancis.com

ISBN: 978-1-138-72322-1 (Pbk)
ISBN: 978-1-315-19313-7 (eBook)

Dedication

Sonia Molly Joseph

Table of contents

About the Author

My name is Paul G. Joseph. In 1983 I graduated from a five year under-graduate program in Civil Engineering at Engineering College, Guindy, University of Madras, Madras, India. I then went on to do graduate work in soil mechanics/geotechnical engineering at Purdue University, Indiana and after that, at the Massachusetts Institute of Technology, Massachusetts. In 2010, I finished a Masters in Applied Mathematics at the University of Massachusetts (Lowell), and in Fall 2013, obtained my Ph.D. I am a registered Professional Engineer (PE) in the State of Massachusetts.

Steady states are ubiquitous in nature and a mathematical framework (loosely called "dynamical systems theory") exists to describe systems with a steady state. The Great Red Spot on Jupiter is an example of a steady state generated by a dynamical system; mathematicians have extensively studied such dynamical systems.

In 1971, Steve Poulos at Harvard first described the steady-state condition in soils. Based on this I was able to show that soil shear can be described as a "dynamical system" whose underlying basis is nothing but Poisson process based simple friction.

These basic findings (steady-state, dynamical systems, Poisson process based simple friction) mark the advent of a new paradigm for describing soil deformation that is at once both simple and powerful. I call this new paradigm Dynamical Systems Soil Mechanics (DSSM for short). It is the only theory that predicts key relationships observed in the empirical evidence from decades of soil tests, relationships which hitherto, have simply been taken as "given."

Looking back on my journey in soil mechanics, it seems nothing but the same old story–The Three Metamorphoses of the Spirit–that Nietzsche powerfully described in his classic Thus Spake Zarathustra. Thus during my undergraduate with Prof. Ramaswamy at Madras University, and then in graduate school at Purdue and MIT, I was like a camel, loading myself with knowledge. After MIT I wandered in the desert until at GEI, thanks to interactions with Gonzalo Castro and Steve Poulos, I turned into a lion. For the next ten years I fought with the great dragon: *"'Thou-shalt,' is the great dragon called. But the spirit of the lion saith, 'I will'."* And then, in the course of this struggle, I became like a child in soil mechanics: *"Innocence is the child, and forgetfulness, a new beginning, a game, a self-rolling wheel, a first movement, a holy Yea."*

I hope you do read this short but rigorous treatment of DSSM–the introduction provides guidelines on how to get the most out of it with relatively little effort. I greatly enjoy receiving questions and comments so please continue to send these to me using the "Send a message" link on the author page for this book (https://www.crcpress.com/authors/i15908-paul-joseph).

Sincerely

Paul G. Joseph Ph.D., P.E.

List of figures

List of tables

Notations

This book uses the following symbols:

A, L and J, K = proportionality constants of model–the initial measures of the resistance offered by the soil structure to q and $\bar{p}$ respectively (dimensionless);

D, B = the exponential rates at which these proportionality constants decay with strain to zero (per unit shear strain, i.e., dimensionless);

C_c = compression index; the change in void ratio per log cycle change in normal effective stress in one-dimensional (1D) compression (dimensionless);

CR = compression ratio; the change in strain per log cycle change in normal effective stress in one-dimensional (1D) compression (dimensionless);

$CSSM$ = Critical State Soil Mechanics;

$C_\alpha e$ = creep coefficient in terms of the change in void ratio per log cycle of time in one-dimensional (1D) compression (dimensionless);

$C_\alpha \varepsilon$ = creep coefficient in terms of the change in strain per log cycle of time in one-dimensional (1D) compression (dimensionless);

$DSSM$ = Dynamical Systems-based soil mechanics;

EOP = End Of Primary, the e-log σ_v one dimensional compression curve where e is the void ratio at the end of primary consolidation;

H = drainage length (dimensionless);

J_e = change in void ratio attributable to the movement of particles into the steady-state or final configuration per unit stress, per unit area of the sample (dimensionless);

J_q, J_p = initial nondimensional interparticle contact area of particles in the shear zone that are not in the steady state or final position, corresponding to q and $\bar{p}$ respectively (dimensionless);

K = ratio of lateral to vertical stress (dimensionless);

K_0 = ratio of lateral to vertical stress for normally consolidated soil (dimensionless);

K_q, K_p, KS_u = nondimensional interparticle contact area of the particles that are not in the flow structure corresponding to q, $\bar{p}$, and S_u respectively (dimensionless);

N_0 = initial number of independent, identically distributed, non-oriented particles, corresponding to some initial stress condition (dimensionless);

N_λ, n = number of independent, identically distributed, non-oriented particles at shear strain γ (dimensionless);

NC = normally consolidated;

OCR = overconsolidation ratio (dimensionless);

$\bar{p}_c$ = original consolidation stress (MLT^{-2});

A_o = original sample cross-sectional area uncorrected for bulging (L^2);

A_c = sample cross-sectional area corrected for bulging (L^2);

$R = d\gamma/dt$ = strain rate ($1/T$); also, number of radioactive particles remaining in the radioactive material;

S_u = undrained peak shear stress (MLT^{-2});

SSL = steady state line; the unique line in q-e space joining steady state points (q_{ss}, e_{ss}) obtained from shear tests run at the same pseudo-static strain rate;

X = strain for which a particle is not in the steady-state flow structure or final structure–a random variable (dimensionless);

b = stress ratio defined as $(\bar{\sigma}_2 - \bar{\sigma}_3)/(\bar{\sigma}_1 - \bar{\sigma}_3)$ (dimensionless);

e = void ratio; Euler's number–the base of the natural logarithm (dimensionless);

e_0 = void ratio at start of test (dimensionless);

h = increment of shear strain (dimensionless);

k = proportionality constant ($1/T$);

u = pore-water pressure (MLT^{-2});

$\bar{p}$, p' = effective normal or confining stress, defined as $(\bar{\sigma}_\alpha + \bar{\sigma}_\gamma)/2$ for a triaxial test and as octahedral confining stress $1/3(\bar{\sigma}_1 + \bar{\sigma}_2 + \bar{\sigma}_3)$ for a true triaxial test (MLT^{-2});

q = shear stress, defined as $(\bar{\sigma}_\alpha - \bar{\sigma}_\gamma)/2$ for a triaxial test and as octahedral shear stress $1/3\sqrt{[(\bar{\sigma}_1 - \bar{\sigma}_2)^2 + (\bar{\sigma}_2 - \bar{\sigma}_3)^2 + (\bar{\sigma}_3 - \bar{\sigma}_1)^2]}$ for a true triaxial test (MLT^{-2});

q_0 = initial shear stress at $\gamma = (MLT^{-2})$;

r^2 = coefficient of correlation (dimensionless);

ss = steady-state;

t = time since start of shear in a constant-strain-rate test (T);

u = void ratio (dimensionless);

α_q, α_p, α_e = secant friction angles corresponding to q, $\bar{p}$, and e respectively (degrees);

$\dot{\delta}$ = rate of peripheral rotational movement in a rotational shear test (LT^{-1});

α = a constant (dimensionless);

ε, ε_{vp} = vertical plastic strain in 1D compression (%);

ε_a = axial strain (%);

ε_1, ε_2, ε_3 = maximum, intermediate, and minimum principal strains respectively (%);

$\dot{\varepsilon} =$ strain rate (1/T);

γ, $\gamma_i =$ shear strain (%), equal to ε_1 in an undrained triaxial test, $\varepsilon_1 - \varepsilon_3$ in a drained triaxial stress where ε_3 is calculated using the volumetric strain and assuming are $\varepsilon_2 = \varepsilon_3$ equal, octahedral shear strain $\sqrt{2}/3\varepsilon_1$ in a one-dimensional consolidation test and the octahedral shear strain $2/3\sqrt{[(\varepsilon_1 - \varepsilon_2)^2 + (\varepsilon_2 - \varepsilon_3)^2 + (\varepsilon_3 - \varepsilon_1)^2]}$ for a true triaxial test;

$\dot{\gamma} =$ shear strain rate (1/T);

$\gamma_s =$ shear strain at the steady-state condition (%);

$\lambda_q, \lambda_p =$ rate of movement of particles into the final state corresponding to q and $\bar{p}$, respectively, per unit of shear strain (dimensionless);

σ and $\sigma' =$ effective stresses (MLT^{-2});

$\bar{\sigma}_1 =$ major effective principal stress (MLT^{-2});

$\bar{\sigma}_2 =$ intermediate principal stress (MLT^{-2});

$\bar{\sigma}_3 =$ minor principal stress (MLT^{-2});

$\bar{\sigma}_a =$ effective axial stress (MLT^{-2});

$\bar{\sigma}_c =$ effective confining stress (MLT^{-2});

$\bar{\sigma}_h =$ effective horizontal stress (MLT^{-2})

$\bar{\sigma}_p =$ preconsolidation stress (MLT^{-2});

$\bar{\sigma}_{p0} = \bar{\sigma}_p =$ preconsolidation stress corresponding to a strain rate of $1 \times 10^-$ (MLT^-);

$\bar{\sigma}_v =$ effective vertical stress (MLT^{-2});

$\bar{\sigma}_{v0} = \bar{\sigma}_v =$ vertical effective consolidation stress corresponding to a strain rate of 1×10^{-7} (MLT^{-2});

$\bar{\sigma}_r =$ effective radial stress (MLT^{-2});

$\bar{\sigma}_{vc} =$ effective vertical consolidation stress (MLT^{-2});

$\tau_{ss} =$ rotational steady-state shear stress (MLT^{-2});

$\omega_{1,2} =$ roots of the differential equation (dimensionless);

Dr. Steve J. Poulos (March 20, 1933–October 6, 2016)

Chapter 1

Introduction

SOIL DEFORMATION IS A POISSON PROCESS: PARTICLES MOVE TO THEIR FINAL POSITION AT RANDOM SHEAR STRAINS

For those new to Dynamical Systems based Soil Mechanics (henceforth called DSSM), the statement above may appear very strange ... difficult to understand, opaque. Hopefully after reading this book it will not appear so; instead, it may now seem to be obvious. As Isaac Newton said over 500 years ago, "Nature is simple, and always consonant to itself." So also, is DSSM (as it should be, seeking as it does, to explain a part of nature). Should confusion set in at any point reading the book, simply take a break, and then come back. DSSM is a very simple theory, needing the knowledge of only a few, basic concepts—simple friction, a Poisson process (radioactivity decay is an example of a Poisson process), and the basics of a kind of mathematics called dynamical systems). This book will systematically uncover these topics.[1]

A Poisson process is a stochastic process that assumes that the interval between each pair of consecutive events is independent of any other event interval. The process gets its name from the French mathematician Siméon-Denis Poisson. Poisson processes can accurately model many phenomena; well-known examples are radioactive decay (Cannizzaro *et al.*, 1978), requests for a document on a web server (Arlitt *et al.*, 1997), and telephone calls (Willkomm *et al.*, 2009). This book describes how soil deformation is also a Poisson process and how, this Poisson process describes the

---

[1] I have tried to write this book as simply and as clearly as possible. However, of course, all authors feel this way! My guideline has been this quote from my favorite philosopher, Nietzsche from his book Beyond Good and Evil: "Being deep and appearing deep—whoever knows she is deep strives for clarity; whoever would appear deep to the crowd, strives for obscurity. For the crowd considers anything deep if only it cannot see to the bottom: the crowd is so timid and afraid of going into the water."

behavior at the particle level of soils, or for that matter, any material made up of particles.

Simple theories that work are powerful theories and so also, here—the six words "soil deformation is a Poisson process" hold a tremendous amount of explanatory power. This book unpacks these explanations systematically, chapter by chapter. Among other things DSSM explains fundamental concepts in soil mechanics that to date have simply been taken for granted, for example, 1) why stress-strain curves normalize with the initial effective confining stress; 2) why in one-dimensional consolidation the void ratio must vary with the log of the effective vertical stress; and 3) why for one dimensional loading the ratio of the creep value C_α must be proportional to the compression index Cc and why this ratio must be approximately constant for a wide range of soils.

This book combines key research conducted over the past five decades into a single, streamlined, coherent work. The target audience is those persons who already have a basic undergraduate background in mathematics and soil mechanics and who are looking for a way to understand DSSM quickly. Before starting though, examine the title of this book. The first part says "Dynamical Systems" while the second says "Soil Mechanics." There is a third part that is missing–Poisson Process–but which silently exists, underlying the other two concepts.[2] In the interest of short titles, keep this missing "Poisson Process" constantly in mind when reading the book.

First, "Dynamical Systems." A dynamical system is just what it says it is—a system that varies dynamically over time. In 1971, Steve J. Poulos, then an Associate Professor at Harvard working for Prof. Arthur Casagrande, formally defined the steady-state condition in soils (Poulos, 1981). A mathematical framework (loosely called "dynamical systems theory") exists to describe systems with steady states. Mathematicians have studied dynamical systems for at least a hundred years. Dynamical systems are found all over in Nature–the Great Red Spot on Jupiter is an example of a dynamical system; other examples include the eye and the heart.

Geotechnical engineers dealing with seismic issues use the word "dynamic" to denote that inertial forces are involved. For dynamical systems though the word is used completely differently—it is used to indicate that the system that as a whole is evolving over time. In the case of this book, only non-inertial, i.e., static conditions are the subject of study, i.e., the system consists of particles that deform only at pseudo-static velocities, i.e., at velocities where inertial forces play no role, and where the system as a whole, with absolutely

---

[2]The Poisson process is a classic example of "emergence," defined as a process whereby larger entities, patterns, and regularities arise through interactions among smaller or simpler entities that themselves do not show such properties. Soil behavior is a classic example of "emergence"–out of random particle movement at the microscopic level emerges order (exponential decay) at the macroscopic level.

no inertial forces, evolves dynamically, i.e., over time. It is important to get this fundamental difference in the two uses of the word "dynamic" clear before proceeding further.

Also before getting into the details, it is important to be clear that the steady-state is different from the critical-state, and that this difference is not a mere matter of "semantics." There is a fundamental difference between the two, so fundamental that while the steady-state condition supports the theory and accompanying machinery of dynamical systems, the critical-state condition does not.

The steady-state is a very precisely defined condition. Poulos (1981) defined the steady-state as follows: *The steady-state of deformation for any mass of particles is that state in which the mass is continuously deforming at constant volume, constant normal effective stress, constant shear stress, and constant velocity. The steady-state of deformation is achieved only after all particle orientation has reached a statistically steady-state condition and after all particle breakage, if any, is complete, so that the shear stress needed to continue deformation and the velocity of deformation remain constant.*

The "steady-state" is defined strictly as per the definition above—every term in the definition is important and a violation of even a single term in the definition results in a state that does not qualify as a steady-state. The steady-state condition applies to drained as well as undrained shear, and to clays, silts, sands and mixtures in between. In fact, the concept of the steady-state applies to any particulates, including food grains such as wheat for example! In soils, the shear stress and the confining stress interact dynamically, each influencing the other, and resulting in the soil's structure reaching the statistically constant condition of the steady-state. Systems evolve over time to reach the steady-state; in the case of soils, the evolution to the steady-state occurs with shear-strain, which in turn, occurs over time. There is nothing critical about the steady-state condition; in fact, the opposite holds in that it is the most stable state a system can enter into, a universal state that a system evolves to ultimately, regardless of differences in its initial conditions.

The steady-state is similar to but not the same as the "critical state condition" first presented by Roscoe, Schofield, and Wroth (1958) in the following description of critical state: *In a drained test the critical voids ratio state can be defined as the ultimate stress of a sample at which any arbitrary further increment of shear distortion will not result in any change of voids ratio. In any series of drained tests, the set of critical voids ratio points so defined can be expected to lie in or near a line on the drained yield surface. In an undrained test the sample remains at a constant voids ratio, but the effective stress p will alter to bring the sample into an ultimate state such that the particular voids ratio, at which it is compelled to remain during shear, becomes a critical voids ratio. In any series of undrained tests the set of critical voids ratio points so defined can be expected to lie in or near a line (not necessarily the same as above) on the undrained yield surface. If the results of drained and undrained tests show that there is, in fact, one unique line to which all loading paths in (p, e, q) space converge, then this will be called the critical voids ratio line (CVR line, for sort). The drained and undrained yield surfaces will then coincide along the CVR line. These two surfaces may be identical and, if so, the common surface would then form the yield surface.*

One difference between the steady-state line and the critical voids ratio (CVR) line as envisioned in this quote is the explicit requirement that the steady-state does not exist unless deformation occurs at constant velocity and with a statistically constant structure that includes the possibility of orientation of the soil grains. In contrast, the implication of the description of critical state in the case of Roscoe, Schofield, and Wroth above is that, if, we stop shear after the specimen reaches the critical state then the specimen remains in the critical state. These authors did extend Arthur Casagrande's original contribution to clays. However, they did not think in terms of a steady-state structure of oriented particles for clays, but rather considered the structure on the CVR to be random.

Recall that at the steady-state the mass of particles is continuously deforming at constant shear stress, normal effective stress, volume, velocity of deformation and with a statistically constant structure. This differs from the critical-state which does not specify any restrictions regarding either deformation velocity or structure. These omissions make it fundamentally untenable to apply dynamical systems theory to the critical-state.

Conceptually, the concept of DSSM is very simple—the shear stress destroys soil structure while the confining stress preserves it; each of these stresses dynamically influence each other, changing soil structure until all the particles that play a role in handling the stresses, reach their final positions. What is happening at the particle level is that under the influence of the forces acting on them, particles move to their final position at random shear-strains and once in their final position remain by and large (statistically) in their final position, at which point their ability to support forces does not further change. Since particles move at random shear strains, this is identical to a Poisson process—a good analog to keep in mind is the Poisson process of radioactive decay where radioactive particles leave the radioactive mass at random times. By definition, this results in an exponential decay of the number of particles that remain in the radioactive mass. The next chapter shows that as with radioactivity, so too with soils.

A reasonably good book on dynamical systems is *Nonlinear Dynamics and Chaos* by Strogatz (1994). Another book, written for the general reader, which became a New York Times best seller, is the book *Chaos: The making of a new science* by James Gleick (1987). In any case, though, one does not have to read either book to understand the soil mechanics dynamical system, simply because the soil mechanics dynamical system is an extremely simple one. Nevertheless, there is one dynamical system that it is important to understand before starting and that is the system that models "big sharks and little fishes," the world famous (among applied mathematicians) Lotka-Volterra dynamical system model.

As described in Brauer and Bies (2011), in the 1920's Vito Volterra, an Italian mathematician, had to model fish and shark populations in the Adriatic Sea. He published his resulting model in 1926. However, researchers later discovered that in the US, in 1910, another mathematician, Alfred James Lotka, had independently created the same model in a different context. Though sometimes called Volterra-Lotka, many use the version with Lotka first simply because he discovered it originally. There are many textbooks and online references to this model. A good place to start (good in that it gives physical meanings for the terms) is Fullford *et al.* (1997).[3]

The Lotka-Volterra model describes how predators (sharks) and prey (fish) interact dynamically and is one of the earliest useful dynamical system models. It opened a whole new perspective to environmentalists as to how simply destroying a predator can result in surprising and non-intuitive outcomes. In a way, at a very high level and very loosely, it is possible to conceive

---

[3] I used this book to learn the basics of dynamical systems. The model it presented for diabetes was a revelation to me for various reasons, including that I did not know medical doctors knew this much math! And this is why I did not include another appendix describing the Lotka-Volterra model–because when you pick up any good book on dynamical systems to read about this model, you will not help but see the tremendous scope and power of dynamical systems theory. Also, a confession: When I first saw the Lotka-Volterra model (in another book) in 2001, I confess that I could not make out head or tail of it. I was not able to connect the symbols of the model to the actual physical phenomena. I stared and stared at it on and off for a couple of days, looking at it for a few minutes, then getting puzzled, and putting it aside. Then one evening I looked at it again and suddenly the light dawned–I suddenly could connect the symbols and the equations directly to the phenomena described. Yes, I know you must think me stupid to take this long to understand this model, I have no excuses other than it had been years since I had done any mathematics, and I had no one to explain all this newfangled dynamical system "nonsense" to me. Later I found that no less than Isaac Newton had said the following when asked how he learned what it took to make his insights: 'I keep the subject constantly before me and wait 'till the first dawning's open slowly, by little and little, into a full and clear light.' Obviously I am no Newton, but I took and continue to take inspiration from his words. Based on my own experience when I enter a new field or consider a new problem, I find that Newton was spot-on when he said this! Once I understood the physical basis of the Lotka-Volterra mathematical model, I knew where I wanted to go with my soils model. The other valuable thing I learned from Newton was this: the French thinker Voltaire visited England and reported that once, when asked how he discovered gravity, Newton did not mention any story of an apple. Instead, he said "... by thinking on it all the time." Likewise, with me too I must say. Most of the insights you will read about here came to me when in the shower or when walking my dog in the woods, or driving my car on some errand, completely lost in thought on matters of soil mechanics. Yes, I know ... I have a dull life. Newton's words "... into a full and clear light" remind me of another great human being–Michelangelo who claimed to be able to "... see the statue in the block of stone."

the soils dynamical system also as a predator (shear stress trying to destroy soil structure) and a prey (the confining stress that seeks to preserve soil structure) model.

The Lotka-Volterra dynamical system is significantly more sophisticated than the soil dynamical system. However, study it long enough to develop a physical feel for the terms of the equations. A physical feel for a mathematical model is always very helpful. In fact, without a physical feel for the model, one may not fully understand nor appreciate the model itself, or those like it, even if they are simpler, as for example, the soil dynamical system model. A thorough understanding of the Lotka-Volterra dynamical system model, will show how simple, the soil dynamical system model really is.

Having covered the first of the two terms, now introspect on the term "soil mechanics." Soil mechanics is a branch of the same mechanics pioneered by the likes of Galileo and Newton. It is science, a branch of physics. Its application to the real world is Geotechnical Engineering.

Science and engineering are very different. Science, by definition, requires a "falsifiable hypothesis." Experimental data are the evidence (called "empirical" evidence) that either falsifies or does not falsify the hypothesis. The basis of all genuine science is the empirical method–the results of evidence obtained by running experiments, compared against predictions from theory. Recall that Galileo almost went to his death because of his belief in this (then new) method of advancing knowledge based on a falsifiable hypothesis testable by experiment, i.e., by empirical evidence.

Engineering on the other hand is the application of mechanics to real world problems. It often uses "thumb rules" or, to use a more formal word, "heuristics" when confronted with problems whose scientific basis is presently unknown. In such situations, engineers pragmatically make assumptions; based on some limited patterns that they have seen they formulate "thumb rules." It is indeed natural that engineers, when confronted with problems where we lack a scientific understanding of the underlying process, then substitute "thumb rules" or "heuristics" based on patterns that they see, but whose basis they do not yet understand. Ponder this deeply but be very clear: do not confuse "the empirical method of science" with thumb-rules or heuristics. These thumb-rules or heuristics are, at best, stopgaps used by engineers until they scientifically understand the underlying physical processes that result in the heuristic.

This book hews to this purist view of soil mechanics being a science. To be a science means it must have a falsifiable hypothesis. To the extent that a concept does not have a falsifiable hypothesis, it is not a scientific theory. Subjective opinion does not come into science–only one question is important–does the theory lead to a model that predicts the experimental data, i.e., does the empirical evidence confirm or falsify the hypothesis.

As the famous Nobel Prize winning physicist, Richard Feynman[4] said "You may have the most beautiful and elegant theory in the world, but if the model that results from it does not fit the data–then your theory is simply wrong." There is a highly-recommended film snippet taken of him teaching a class at Caltech (Feynman, R., 2016) where he clearly and emphatically makes this point.

In this film, Feynman emphasizes the critical importance of testing theory with experiment and that it does not matter if one's theory is "beautiful" if it does not match the empirical evidence.[5] Never forget this! In addition, while on this subject of empirical evidence, i.e., test data–if someone writes "… the fit was good" but then fails to provide numerical measures of the fit, what they are doing is really "chi by eye."[6] This is a phrase used derogatorily by scientists for those who judge how good a model fits its data, purely on a visual basis and without providing goodness of fit ("chi") values. Be sure that if presenting a paper at a conference you include goodness of fit information, else, someone in your audience may call you to task.

One more thing–when reading the source paper for a constitutive model, it is typical to find that the authors present matches to only two or three stress-strain curves of one soil type–typically an insensitive clay, i.e., a clay whose stress-strain behavior is very simple, with little or no strain-softening. This does not meet standard expectations of a satisfactory attempt at falsification with empirical evidence. What these authors have done is to pick out of all their tests, the few that came closest to their model and to use these few to justify their model.

This method of testing is "conformal testing" and is very different from "falsification testing"–testing whose specific aim is to falsify the hypothesis, and which drives good science. One famous scientist who adhered strictly

---

[4]It is interesting that Feynman had a relatively modest IQ of 125. This confirms my bias against IQ tests–that it is more a measure of one's analytical abilities and that it says nothing of one's creative abilities. Creative abilities depend on a strong power of imagination, grit, passion, and a fundamental openness to new ideas–and IQ tests do not measure these characteristics.

[5]Usually, credit goes to Roger Bacon (c. 1214–1294 ACE) for introducing the modern scientific method, i.e., the practice of validating hypothesis with experiment. Until then, the basis of "science" was Aristotelian logic, which had almost no concept of experimental validation. For example, a common opinion that stood untested until Galileo's time was that heavier objects fell faster than lighter ones. If you lived in the 14th century and asked a scholar learned in Aristotelian logic about falling objects, he would look you gravely in the eye and assure you that heavier objects fell faster than lighter ones. Persons like these who pay no heed to experimental data and its need to match theory are "Aristotelian Scholastics." The important thing to take away is that if one's opinion does not agree with the theory, but if predictions from the theory match the data, one should question one's opinion as much as or more so than the theory.

[6]Americans pronounce "chi" to rhyme with "eye."

to this concept was Newton who was known for obsessively trying to prove his own theories wrong. The research described in this book raises the bar, by trying to falsify the model across many test conditions!

A constitutive model must demonstrably prove its ability to match data from at least over a hundred of tests, on soils of different types (sands, silts, clays and combinations thereof), and under various conditions of drainage, over-consolidation ratio, and shear, including most importantly, the strain-softening behavior seen in many clays. Then too it is important to compare the model's outcome against the test data to at least 25% axial strain and not, as is often the case, simply stop at somewhere less than 10% (because after this, the model does a terrible job of matching the measured data). It is also essential that the model well fit void-ratio or pore pressure changes with shear. This latter match is particularly difficult to achieve, i.e., is a good way to falsify a model, and hence, important to see in any given model.

The entire subtext of this book is a constant effort to show that soil-mechanics is a "science" and is no longer "engineering." In short, a key thrust of DSSM (and also of this book) is to bring soil mechanics fully under the umbrella of standard science, including, researching it as one. Its application to the real world is "geotechnical engineering." The concluding chapter of this book on Finite Element Analysis explains how the science of DSSM transitions into engineering, with its application.

The book consists of four parts–one on preliminaries (this chapter), a second on shear, a third on consolidation and creep, and a fourth on applying the model to solve real world problems using Finite Element Analysis. The second part on shear consists of the Chapters 2 through 5. Of these, Chapter 2 deals with the phenomenological model, a model that describes observed shear behavior, i.e., the phenomena of soil shear, without trying to model the physical processes that underlie and drive such behavior. Chapter 3, the key chapter of this book, describes the underlying physical bases of this phenomenological model, Chapter 4 the implications of this model, and Chapter 5 highlights the strain-rate behavior of soils in shear, in the context of dynamical systems theory.

The third part has two chapters that deal with one-dimensional compression. The first of these, Chapter 6, generalizes the dynamical systems soil theory and shows how to use it to derive the well-known linear relationship between void ratio and the log of the effective vertical stress seen in one-dimensional consolidation. At present, DSSM is the only theory that can derive this relationship from fundamentals, and not as do other theories, simply, take it as a given.

The second chapter in this section, Chapter 7, deals with creep and viscous effects in one-dimensional compression loading. Again, we see that DSSM theory transparently and easily explains another relationship simply taken to date as simply a "given" and that is why the ratio of the creep coefficient to the primary compression index is a constant. This chapter also resolves a long-standing controversy on the role of viscous effects in one-dimensional loading. This is the power of simple theories that get to the root cause of behavior and lays bare its underlying physical mechanisms. Such "physically based theories" pack tremendous explanatory power, answering questions that simplistic examination of mere outward behavior, i.e. phenomena, cannot.

The fourth and final part (Chapter 8) shows how to use DSSM in Finite Element Analysis.

A typical undergraduate student in the US should take a day or two to understand and develop a physical feel for the Lotka-Volterra model and an hour for this introduction. The remaining chapters will take about a day or two each. Then also, first read only Chapters 2, 3 and 6. These three chapters, particularly Chapter 3, explain the core concepts behind DSSM.

The remaining chapters (4, 5, 7, and 8) are for graduate students and researchers or those otherwise very interested in this subject. Chapter 8 is only for the most mathematically prepared. Likewise, the Appendices on the derivation of the Poisson process as it applies to soils and the approximate analytical solution to the DSSM model are only for the most mathematically prepared. If keen on learning about "emergence" and why soil behavior is a classic example of emergence then read footnote 2 in Appendix 1.

In any case though, regardless of how long it really takes, fully understand each chapter before proceeding to the next. For example, try not to go to Chapter 3 until able to picture in the "mind's eye," the dynamical system of Chapter 2. This concept of the "mind's eye" is important–it is the ability to visualize mentally, in this case, the model and its underlying mechanisms.

Neuroscience suggests that engineers, scientists, and mathematicians, regardless of whether they have a male or female brain (or a brain that is in-between) are far better at mental visualizing than the average population. This makes sense because for the brain too, as elsewhere, practice makes perfect! So even if at the beginning, it may be hard to do this kind of visualization, the common experience is that with practice, it becomes easier and easier, until finally it becomes automatic and second nature.

It helps also to write down the key concepts, using pencil and paper. The tactile feel of pencil on paper goes a long way to getting one's very physical body to seamless interact with the brain and channel the information in.[7] Then try to verbally explain the concept with books closed. About 2,500 years ago, Aristotle said, "If you cannot say it, then you do not know it." (He was not talking though about reciting the material, parrot like, from memory and with no understanding!) Some chapters are quite mathematical—after

---

[7]Note that this falsifiable hypothesis now seems to have some empirical evidence behind it (see Muller and Openheimer, 2014).

reading one such chapter, close the book and try to derive the equations presented in the chapter.

Notice that the book repeats important concepts using different words. This is intentional so that by meeting them more than once, they will be "ever present" and it will soon become second nature to draw upon them when thinking about soil mechanics.[8]

One of Nietzsche's aphorisms is this one: "The doer alone learneth." To really learn something thoroughly, first throw away the book then, go to work with the concepts gleaned from them. If however disaster ensues, quickly retrieve the book and clarify the material in question; then throw the book away once more and try again! Soon, once all its information has been "internalized," there will be no need to retrieve the book, even when tackling the most complex problems. What has happened is not mysterious– by dint of hard work, i.e., by the old-fashioned way, one now thoroughly understands the subject material!

After reading the book then, become one of Nietzsche's "doers" and "bake it in deeper" by applying the model to some test data, or by teaching someone else DSSM, or by using DSSM to figure out why some soil behavior is the way it is. Teach a course on DSSM to undergraduates or peers. Remember this statement by Einstein always though: "If you can't explain it simply, you don't understand it well enough.[9]"

For undergraduates, Chapters 2, 3, and 6 should suffice to understand the core concepts of DSSM. The remaining chapters are for advanced undergraduates, graduate students and researchers.

To summarize, few basic, standard concepts from science (the steady-state condition, dynamical systems theory, simple friction and the Poisson process) mark the advent of a new paradigm for describing soil deformation. This new paradigm is at once both simple and powerful and because it is falsifiable at many levels but remains unfalsified despite various attempts to falsify it, counts as a scientific theory. This new paradigm is Dynamical Systems Soil Mechanics (DSSM for short).

---

[8]I read of this technique in the introduction to the book *Principles of Soil Mechanics* (Scott, 1963).

[9]*Einstein: the life and times* pp. 343 by Clark (1971): Louis de Broglie did attribute a similar statement to Einstein. To de Broglie, Einstein revealed an instinctive reason for his inability to accept the purely statistical interpretation of wave mechanics. It was a reason that linked him with the physicist Rutherford, who used to state, "it should be possible to explain the laws of physics to a barmaid." (I personally think that Rutherford was snooty, condescending, and sexist.) Einstein, having a final discussion with de Broglie on the platform of the Gare du Nord in Paris, whence they had traveled from Brussels to attend the Fresnel centenary celebrations, said, "that all physical theories, their mathematical expressions apart ought to lend themselves to so simple a description 'that even a child could understand them.'"

It is the only theory at present that explains relationships in soil mechanics hitherto simply taken for granted. These include, for example, why there is a linear relationship between the void-ratio and the logarithm of the vertical effective stress for the case of one-dimensional compression. Other theories, simply take such relationships as "given" and do not even try to derive them from first principles as is possible with DSSM. Go through the rest of this book and introspect deeply on it.

Dr. Steve J. Poulos constantly emphasized his and Casagrande's belief in the importance of handling real soils physically, using one's hands. Steve passed away on October 6th, 2016, aged 83. Sadly, during his lifetime, the soil mechanics community at large never formally recognized him for his seminal work in soil mechanics. The concept of DSSM is an extension of research on the steady-state by Prof. Casagrande and the soil mechanics department at Harvard College including Dr. Steve J. Poulos, Dr. Gonzalo Castro, and Dr. Daniel P. LaGatta.

Chapter 2

Soil shear–the phenomenological model

A phenomenological model is a model that describes observed behavior (phenomena), without trying to model the underlying physical processes that cause or drive such behavior. Often in problems that appear to be complex, the first step in modeling something is simply to model only the observed behavior. Then, with this model in hand, one can dig deeper and search for the likely physical basis of the observed behavior. This chapter describes this initial work modeling the observed behavior of a soil sample when sheared. With the phenomenological model firmly in hand, the next chapter will proceed to determine the root cause of the behavior, i.e., the physical basis of the behavior and use this to create the so-called "physical model."[1]

---

[1]The model described in this chapter is an example of "deductive reasoning," one of the two common methods of the scientific process. In deductive reasoning one goes from the general (the theory) to the specific (logical conclusion or hypothesis), then tests the hypothesis with empirical observations. This chapter is an example of this deductive process–from the general theory of dynamical systems which applies to dissipative systems (a system operating in an environment with which it exchanges energy and/or matter) such as soil, I formulated the specific, logical hypothesis that it entailed, namely that soil deformation must also be a dynamical system. Note that I had by this time through understanding of Steve Poulos's steady-state concept and so this was not a hard leap for me to make! I then developed the specific details of this hypothesis (the equations of the model) and tested the hypothesis using empirical data. The second common method of the scientific process is "induction" where one goes in the reverse direction–from specific observations (usually patterns in the empirical evidence or data) to general theory. The next chapter, which describes the physical model, is an example of induction–I went from the specific pattern of exponential decay that I saw in the triaxial test data to the general theory of simple friction and a Poisson process. Induction is usually easier than deduction and my experience with both leads me to agree! It took me about ten years to deduce the phenomenological model described in this chapter, but only three months to induce the physical model described in the next chapter, a difference in effort of about 40 fold! Some researchers tend to use only one approach because they are "geared" that way. It appears that I have no such preference, seeming to use whichever method seems natural. In fact, I am not even aware of which method I use until well after

However, before starting on this chapter, first thoroughly understand the Lotka-Volterra dynamical system mentioned in the introduction. Visualize the big sharks and little fish and how they interact, given this understanding of the Lotka-Volterra model. Knowledge of the Lotka-Volterra model will make this chapter simple and easy to understand.

When sheared, a soil starts from an initial condition and ultimately reaches the steady-state condition. This book uses the word "steady-state" strictly as per its definition in Poulos (1981): *The steady-state of deformation* for any mass of particles is that state in which the mass is *continuously deforming* at constant volume, constant normal effective stress, constant shear stress, and constant velocity. The steady-state of deformation is achieved only after all particle orientation has reached a statistically steady-state condition and after all particle breakage, if any, is complete, so that the shear stress needed to continue deformation and the velocity of deformation remain constant.

The steady-state condition applies to drained as well as undrained shear, and to clays, silts, sands and mixtures in between. In fact, the concept of the steady-state applies to any particulates, including food grains such as wheat for example![2] The steady-state is a very precisely defined condition. At the steady-state the mass of particles is continuously deforming at constant shear stress, normal effective stress, volume, and velocity of deformation. Various other systems in nature likewise dynamically evolve from some initial condition, to a final steady-state condition. Researchers have routinely used the framework of dynamical systems theory to study such systems (see Strogatz 1994 for examples).

Poulos (1981) describes the two key differences between the steady-state and the critical-state condition. First, the steady-state has a requirement that the deformation explicitly occurs at a constant deformation velocity (which can be any velocity that does not cause inertial effects), and not as in the case of the critical-state, only the near zero velocity. Second, at the steady-state the soil grains position themselves in the steady-state flow structure that remains statistically constant with continued shear strain, one example being needle or plate like particles oriented all in the same direction for clays sheared to the residual condition. No such restrictions apply to the critical-state.[3]

I have finished the work and then, several years later (as at present for example), look back on my findings and recognize in hindsight, which approach I took.

[2] I think I understand why Steve mentioned food-grains. If you travel even once in the mid-West in the US, you will come across huge cylindrical towers that hold wheat. These grain storage bins have a certain fascination for some soils engineers in the US because unexpected behaviors occur when storing grains in huge quantities. For example, on one occasion (when I was studying at Purdue), tragically, young farmer accidentally fell into one of these bins and "drowned" in the grains of wheat. Grains can behave like a fluid under some conditions of packing.

[3] One should not re-label the steady-state as the critical state! This would be an egregious, fundamental mistake for two simple, fundamental reasons. First, the goal of

These differences between the steady-state and the critical-state are not a matter of semantics but have deep implications. They allow for the application to the steady-state concept of soils, the mathematical framework commonly known as "dynamical systems." The concept of dynamical systems cannot apply to the critical-state where deformation occurs at the pseudo-static velocity and without no specifications as to soil structure.

The steady-state of deformation occurs in a particulate mass for any loading and drainage condition where the shear stress breaks down the original structure and puts it into a new "flow" structure. It is the result of a continuous process that starts with the first small increase of shear stress and ends when the soil reaches a steady-state of deformation, with all trace of the original structure lost, and with a statistically constant structure.

Soil particles move and reorient continuously during shearing and even in the steady-state flow-structure, for elongated particles, there can be small, continual, random movements, while for bulky grains, there can be grain rotations. However, the steady-state flow-structure is statistically constant in that the movements of these individual particles result in negligible changes in the shear stress, normal effective stress, volume (void-ratio), and velocity observed with continued straining at the steady-state (Poulos, 1981, 2010). The well-known residual condition resulting from large strains of clays is an example of a steady-state condition. The steady-state applies to clays, silts, and sands, and combinations thereof.[4]

---

science is to unify disparate fields together and to explain them with minimal theory. Dynamical systems based applied mathematics is an established field of mathematical research, one that is at least a hundred years old, and which has defined terms that are well-known and understood. The steady-state condition is one of these well-defined terms, one used for systems very far afield from soil mechanics in terms of phenomena, such as astronomy and space-physics (for an early example, see Sciama, 1955). Second, and of fundamental importance, is that there is nothing "critical" about the steady-state condition. In fact, it is the least critical of all conditions a soil can be, because it is always the end condition, regardless of any initial conditions, however disparate these initial conditions may be. For these reasons, to re-label the steady-state as the critical-state is a fatal, egregious mistake because it is needless, confusing, deeply against scientific mores, and also, a misnomer.

[4]One very important thing that I learned during this research program was that thermodynamically, soil deformation is a dissipative process and that there are proofs to show that all thermodynamic dissipative systems must have steady-states (see for example, Nicolis and Prigogine, 1977). Indeed, were soil deformation not to possess a steady-state, it would violate the fundamental laws of thermodynamics, falsifying them and forcing a reconsideration of the entire field of thermodynamics. This is very unlikely to be the case. This is why I am tired of naive understandings of the steady-state condition in soil, the most common one being that only soils with "needle or plate like" particles can "align" to a steady-state. These understandings are naive—the steady-state is a statistically constant structure, and even an assemblage of perfectly spherical ball-bearings made of steel and produced to extremely narrow tolerances, can reach steady-state conditions. Of course, a "statistically constant structure" is

The Steady-state Line (SSL) is the line joining the steady-state values of shear stress, plotted against the steady-state void ratio. It is unique for a given strain rate, as expected from the definition of the steady-state as a unique combination of steady-state shear and confining stresses, and void-ratio. On the other hand, the critical-state condition's lack of precise definition both in terms of structure, and strain-rate results in different experimenters interpreting it differently, both across test programs, and sometimes by even

---

something to stop and ponder about. As we know from the introduction, science is fundamentally about attempting to falsify hypotheses. As you will see in the next chapter, the test data for sands show that they match the dynamical-systems model, a model whose fundamental assumption is that the structure is heading to the steady-state. In other words, leaving mere "subjective opinion" behind, and applying the model to data, we find we are not able to falsify the hypothesis that rounded or sub-rounded particles have a steady-state. A statistical constant structure means that individual particles could well be rotating on one of multiple axes, but that this rotation of individual particles need have no effect on the soils overall void-ratio, effective confining stress, and the structure's ability to resist shear-stress at that confining stress. A few individual particles could even be moving to new configurations but with other particles compensating so that there is no net effect in the aggregate (remember that soils are but aggregates of particles and that soil behavior is but the behavior of an aggregate of particles and does not depend on what a given particle is doing at a given instant of time). In short, one needs a "physical feel" for all types of soils so as not to be confined to thinking that the steady state concept applies merely to soils with needle or plate like particles. One of my teachers at Purdue, Gerald Leonards, once told me in 1984, that a physical feel for a phenomenon comes only with deep introspection. Additionally, I think that one obtains this "physical feel" in a quite literal sense only after one has extensively interacted physically with the object of one's introspection, using one's hands. Limited "hands-on" interaction is far from enough to develop the physical feel needed for soil mechanics research, and leads to dangerously flawed idealizations of soils behavior. And no, neither "geotechnical consulting" (so called) nor "geotechnical engineering design" can give you the "raw feel" for soils that I am talking about. A New York Times non-fiction best-seller that captures this view point exactly is: Shopcraft as Soulcraft: An Inquiry into the Value of Work (Crawford, 2010). I think it is essential reading if you want to become really good in soil mechanics! When I got out of graduate school, I found myself assigned to a soils laboratory. I was to remain in a soils lab for almost five continuous years, running plastic limit tests, proctor compaction tests, grain sieve analyses, hydrometers, and various kinds of shear tests. I hated it, and thought my career as an engineer was over. Little did I know that my forced confinement to the laboratory was informing my understanding of soils in a very deep and fundamental way–my "body intelligence" was obtaining information that supplemented what my mind already knew from mere book-study. At my best, a soil became a "live" thing to me. I could simply pick up a handful of any soil and at once it would "talk" to me, telling me among other things its Proctor compaction value, and if it was plastic, its liquid and plastic limit. Steve Poulos, not only strongly felt likewise, but as he told me, so also did Arthur Casagrande. So please before you write me to say that steady-states in soils are only to be found in clays with needle or plate-like structures, first introspect on what a "statistically constant structure" means; next attempt to visualize in your mind's eye, a collection of steel ball-bearings, reaching a steady-state.

the same experimenter within the same test program. This in turn often results in a non-unique critical-state line, particularly for strain-softening tests, a confusion that some researchers carry over to discussions on the uniqueness of the steady-state line.

Earthquake engineering, large-strain, and soil dynamics research use the steady-state concept. (See for example: Lade and Yamamuro (2011), Fourie and Tshabalala (2005), Okada *et al.* (2005), Wang and Sassa (2002), Yamamuro and Covert (2001), Yamamuro and Lade (1998), Riemer and Seed (1997), Finno *et al.* (1996).) Joseph (2009, 2010) describes the phenomenological soil shear model covered in this chapter.[5] The model is based on the steady-state concept, and consequently, allows for modeling soil stress-strain curves through the entire shear strain range, and not just till when shear failure planes develop.

Joseph (2009, 2010) proposed that for monotonic shear tests, the rates of change of shear stress q, effective normal stress $\bar{p}$, and void ratio e are proportional to the applied shear and effective normal stresses, with the initial proportionality values decaying with shear strain γ to reach ultimately a value of zero at the steady-state condition. Following other systems in nature, the proportionality values decay exponentially, as the soil structure changes under shear from its initial structure, to its final steady-state flow-structure. The decay is a result of the soil structure changing with strain from its initial structure to its final steady-state structure. Shear stress drives structural change at a certain rate, while effective normal stress resists it at a different rate.[6]

---

[5] I cover the details from all my papers in this book. You do not need to obtain or read my original papers unless of course you very much want to, because all the material in these papers (and more) is contained right here. Further more, I took the opportunity when writing the book to correct errors that had crept into some of the papers and to word things more clearly. In short, this book should be seen as superseding any paper of mine that it quotes.

[6] Note that these statements consist of a set of hypotheses that are falsifiable. Either the resulting model will match the test data or it will not–no excuses. In Equations 1, 'e' is not the void-ratio. Rather, it is the Euler number, the base of the natural logarithm. Later on, to avoid confusing it with 'e' the void-ratio, I am forced to use '*exp*' instead, a clunky borrowing from (originally) the FORTRAN representation of 'e'. The number e is one of the most important numbers in mathematics. It is an irrational, transcendental number whose first few digits are: 2.71828182845904523535602874713527... (and forever more). To have to use the clunky 'EXP' instead in calculations is needless (why not instead use a simple, direct, and intuitive 'v' for void-ratio?). Using 'e' for the void-ratio signifies a basic ignorance of and insensitivity towards mathematical niceties and in a future edition of this book, I may switch to using 'v' as the symbol for the void-ratio. Anyway, curious about this abomination, I asked Dick Goodman, the author of a noted biography on Karl Terzaghi (Goodman, 1998), what its origin was. He wrote to me that: '... *in Erdbaumechanik, (1925) Terzaghi used the Greek letter epsillon (ϵ) for the pore number (Porenziffern), which was identified as "the quotient of the volume of voids to*

The physical behavior described above expressed mathematically is:

$$\frac{dq}{d\gamma} = \bar{p}Ae^{-B\gamma} - qJe^{-D\gamma} \tag{1a}$$

$$\frac{d\bar{p}}{d\gamma} = \bar{p}Le^{-B\gamma} - qKe^{-D\gamma} \tag{1b}$$

where: A, L, and J, K are the proportionality constants and stand for the initial measures of the resistance offered by the soil structure to $\bar{p}$ and q respectively; D, B are the exponential rates at which these proportionality constants decay with strain to zero.[7] Note that the model does not describe initial elastic deformation, and that parameters A, J, L, K, D, and B are all strain-rate independent. Equations 1 satisfy the steady-state requirement of zero change at the ultimate condition. Note also that Equations 1 have no error correction terms. The model neither describes elastic strains nor applies to cemented soils, meta-stable soils, or soils with significant particle breakage on shearing. At present, it applies only to monotonic shear tests.

The model described by Equations 1 fit closely when applied to stress-strain curves – Figure 2.1 below shows two examples for two tests from Sheahan (1991) on Boston Blue Clay. These tests have nothing in common other than the goodness of fit of Equations 1 above are typical for their test type. (The next chapter details how to calibrate the model to test data as shown in Figure 2.1, in an Excel spreadsheet, using the Range Kutta method.)

Observe how in Equations 1a if the sample is subject to only pure hydrostatic stress, then, if the soil has an anisotropic structure (visualize a house of cards like what children sometimes build), it follows that there will be shear

---

the volume of grains". I looked into Terzaghi's 1929 book, Ingenieurgeologie (engineering geology) with Redlich and Kampe, (Redlich et al., 1929) and found he was now using "e" for void ratio. Since in clays he saw this as measure of the distance between particles, I thought he might have selected "e" to represent a special kind of distance — the word "entfernung" comes to mind. It means "distance". But I have no information to suggest that this is the origin of the symbol. Of course, he might just have tired of using a Greek symbol and replaced it with the Latin e ...' In this edition of the book I usually use 'e' to represent void ratio, though occasionally, to highlight the visual elegance of an equation, I use 'e' to stand for base of the natural logarithm, instead of its clunky (FORTRAN) equivalent "exp".

[7] Critical state soil mechanics (CSSM) models that assume zero shear strain and shear stresses for pure hydrostatic compression are making an unconservative, incorrect assumption. It is an assumption made by many CSSM elasto-plastic and Mohr-Coulomb based models because otherwise, various numerical instabilities arise. Nonetheless, it is a dangerous idealization to make.

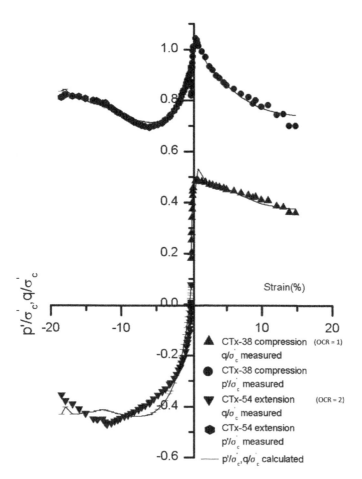

Figure 2.1 Typical measured (Sheahan, 1991) vs. calculated (Eq. 1) normalized stress-strain curves.

strain. In other words, even if shear stress starts out zero, on the first application of a pure hydrostatic stress, as shown in Equation 1a, there will be an increment in shear stress with respect to shear strain. Likewise, Equations 1b show that by applying pure shear stress alone, there will be a change in hydrostatic stress with respect to shear strain.

This models the real behavior of soils and contrasts strongly with the highly artificial assumption routinely made in almost every elasto-plastic constitutive model that only shear stress causes shear strain and only hydrostatic stress causes volumetric strains. This artificial condition only holds if the fabric (including grain-shape) of the soil is completely isotropic. While extreme isotropic structure such as this may occur in yielding metals, it is almost impossible to find this degree of isotropic structure in a real soil.[8]

As mentioned earlier, the physical basis of the model and its parameters remain undefined, i.e., the model describes the observed behavior (phenomena) of soil deformation under shear but not the underlying physical basis of this behavior. The next chapter details the physical basis of the model. However, before continuing to the next chapter visualize shear stress trying to destroy soil structure with confining stress trying to preserve it, with the two interacting till they stabilize. As with the Lotka-Volterra model, try to write the equations for the phenomenological model (Equations 1) without looking, based not on memory, but on deep understanding.

---

[8]To be poetic, in a way Equations 1 are really describing the ancient Chinese principle of yin and yang where yin is analogous to preservation (the confining stress) and yang is analogous to action (the shear force). The other interpretation is Shiv-Shakti, the ancient Indian concept, where Shakti is the female consort of the male Shiva. Shakti symbolizes action and energy (shear), while Shiva symbolizes stability and maintenance (confining stress). The name given to the interaction between the two is *rasa-leela*, which is Sanskrit for "play of life." The similarity of these two ancient concepts with the soil deformation dynamical system has always impressed me profoundly; I think it is no accident. It implies perhaps that this is a general principle found commonly in nature, that the ancient shamans were able to see/intuit.

Chapter 3

Soil shear–the physical model (part 1)

This is the most important chapter in the book. If there is time to study only one chapter in this book, let it be this one. It is written to be self-contained, and is the 20 percent of the book, that describes 80 percent of the theory behind DSSM. Understand it thoroughly to build a solid foundation in DSSM.

Chapter 2 defined the steady-state condition and showed how for mono-tonic shear tests, the rates of change of shear stress, effective normal stress, and void ratio are proportional to the applied shear and effective normal stresses, with the initial proportionality values decaying with shear strain to reach ultimately a value of zero at the steady-state condition. It described the "phenomenological model," i.e., a model that described soil deformation phenomena, but which said nothing about the underlying physical processes that could account for these phenomena.

This chapter goes one-step deeper–it describes the physical basis of the phenomenological model, proposing a stochastic process as the physical basis for the earlier model and its parameters. This physical model, when applied to 130 shear tests from various research programs on uncemented clays, silts, sands and mixtures thereof, under a variety of test conditions, closely fit the data together with an orderly variation of the model parameters. As before, the model neither describes elastic strains nor applies to cemented soils, meta-stable soils, or soils with significant particle breakage on shearing. At present, it applies only to monotonic shear tests.

When reading this chapter, keep in mind that this book strictly adheres to the correct definition of the steady state in particles from Poulos (1981): The steady-state of deformation for any mass of particles is that state in which the mass is continuously deforming at constant volume, constant normal effective stress, constant shear stress, and constant velocity. The steady-state of deformation is achieved only after all particle orientation has reached a statistically steady-state condition and after all particle breakage, if any, is complete, so that the shear stress needed to continue deformation and the velocity of deformation remain constant.

If the deforming mass of particles do not meet even one of the conditions in the definition, then the particles are not in the steady-state. During shear, particles move and reorient continuously and even in the steady-state flow-structure, for elongated particles, there can be small, continual, random movements, while for bulky grains, there can be grain rotations. However, the steady-state flow-structure is statistically constant in that the movements of these individual particles result in negligible changes in the shear stress, normal effective stress, volume (void-ratio), and velocity observed with continued straining at the steady-state (Poulos, 1981, 2010). The well-known residual condition resulting from large strains of clays is one example of a steady-state condition. The steady-state applies to clays, silts, and sands, and combinations thereof.

THE MODEL

Particles (a single particle or cluster of particles) transfer forces through their contacts with other particles. Due to the change in load on the soil, these inter-particle forces change, causing particles to reorient from their initial position to reach eventually the steady-state flow-structure. This change in structure caused by this reorientation in turn changes the ability of the structure to support shear and normal stresses. As particles move into the steady-state flow structure, the number of particles available for reorientation under conditions of ongoing shear strain application decreases with shear strain. If no such reorientation of particles occurs, shear deformation will proceed at the initial value of the shear stress, normal stress, and void ratio, none of which will change with shear strain because they will already be at their steady-state values.

Note that the concept of the steady-state applies in general to all particles and so too do most of the concepts in this book. Since soils constitute an assemblage of particles, the concept of the steady state and its associated theory (this book) also applies to soils. For soils the key, falsifiable hypothesis is that the physical basis of soil deformation is a micro-structural process of particle reorientation with a particle moving into the flow-structure at some random shear strain. This is exactly like the case of radioactive decay where radioactive particles leave the radioactive material at random times. For such processes, the expected value of items that remain (be they radioactive particles or soil particles not in the steady-state position) decays exponentially as[1]:

$$N = N_0 \exp(-\lambda\gamma) \tag{1}$$

[1]Processes such as these Poisson processes, named after the great French mathematician, Siméon Denis Poisson. Indeed, I count Poisson as a true genius–he uncovered an example of nature's sublime simplicity. Think about it–nature puts no restrictions and allows particles to behave freely, randomly ... and out of that, as Poisson discovered, emerges robust order! Exponential decay in this case. The mechanism behind this robust order is so simple as to be almost invisible, needing a keen, keen nose to smell out yes, Einstein credited his "nose" for guiding him towards general relativity though of course he did not mean his real nose!

where for the case of soils, N_0 is the initial number of particles that are not in the steady-state position, and N is the remaining number of particles not in the steady-state at shear strain γ.

Note: the detailed derivation of this equation is in Appendix 1.

As particles move into the flow-structure, most particles will affect both shear stress q and effective normal stress $\bar{p}$, though some particles will affect only q and some only $\bar{p}$, i.e., the numbers of particles whose reorientation affects q will be close but not exactly equal to the number of particles whose reorientation affects $\bar{p}$. Let N_q and N_p be the number of particles that affect q and $\bar{p}$ respectively, and let J_q and J_p be their corresponding inter-particle contact area per unit sample area. Since J_q and J_p are surface areas per unit area of the sample, they are non-dimensional.

As particles orient into the flow-structure under the action of q and $\bar{p}$, they no longer play a role in the change of q and $\bar{p}$, and so the initial values J_q and J_p are subject to a decrease as shear progresses. The rates at which particles orient to q may be close to but not necessarily the same as the rate at which they orient to $\bar{p}$, and consequently, the initial values of J_q and J_p are subject to decrease at correspondingly different rates. Let these rates of decrease be λ_q and λ_p respectively. Following from Equation 1, the interparticle contact area of particles that affect q and not oriented with q and $\bar{p}$ are respectively $J_q\exp(-\lambda_q\gamma)$ and $J_q\exp(-\lambda_p\gamma)$ at any given shear strain γ. Likewise, the inter-particle contact area of particles that affect $\bar{p}$ which not oriented with q and $\bar{p}$ are respectively $J_p\exp(-\lambda_q\gamma)$ and $J_p\exp(-\lambda_p\gamma)$ at any given shear strain γ.

Due to the normal force per unit area of J_q and J_p, the shear resistance is $\bar{p}\tan\alpha_q$ and $\bar{p}\tan\alpha_p$, where α_q and α_p are the secant friction angles of the particles whose reorientation cause changes in q and $\bar{p}$ respectively. Though the underlying material friction angle is the same, the friction angle component due to interlocking will be slightly different for particles whose movement is driven by changes in q versus particles whose movement is driven by changes in $\bar{p}$. This shear resistance from the normal force opposes the shear force acting on surfaces J_q and J_p, and the rate at which q and $\bar{p}$ change is the difference between the two, i.e.:

$$\frac{dq}{d\gamma} = J_q[\bar{p}\tan\alpha_q\exp(-\lambda_p\gamma) - q\exp(-\lambda_q\gamma)] \tag{2a}$$

$$\frac{d\bar{p}}{d\gamma} = J_p[\bar{p}\tan\alpha_p\exp(-\lambda_p\gamma) - q\exp(-\lambda_q\gamma)] \tag{2b}$$

Likewise, for drained shear, the change in void ratio e is:

$$\frac{-de}{d\gamma} = J_e[\bar{p}\tan\alpha_e\exp(-\lambda_p\gamma) - q\exp(-\lambda_q\gamma)] \tag{2c}$$

where:

J_e is the initial void ratio due to non-orientation of particles, per unit area of the sample.

There is no change in the number of parameters from the model proposed in the phenomenological model of the previous chapter. The key change from that earlier model is the replacement of the symbols J, K, M by J_q, J_p, J_e and symbols A, L, N by $J_q \tan \alpha_q$, $J_p \tan \alpha_p$, $J_e \tan \alpha_e$ respectively. The significance of this change is that now the model parameters have a known physical basis. Note that the same single process of particles moving into the steady-state flow-structure at random shear strains causes the changes in q and $\bar{p}$ (and for the drained case, e).

As with the phenomenal model described in Chapter 2, note how for the physical model described in this chapter, if you apply only pure hydrostatic stress, then, if the soil has an anisotropic structure (visualize a house of cards), it follows that there will be shear strain. In other words, even if shear stress starts out zero in the right-hand side of Equations 2a, on the first application of a pure hydrostatic stress, with the advent of shear strain, Equation 2a shows that there will be an increment in shear stress.

The only case where this will not happen is if the soil fabric and soil particles are both completely isotropic. While such a perfectly isotropic configuration may occur in yielding metals, they do not occur in real soils. This is one of the key flaws of most CSSM based constitutive models—they assume zero shear strain and shear stresses for pure hydrostatic compression and zero hydrostatic strain or pore pressures for pure shear. They do this because otherwise, various numerical instabilities arise. Nonetheless, it is a naive, flawed, and dangerous assumption to make.

In the context of a constant strain rate shear test, replacing shear strain γ by Rt and the change in shear strain $d\gamma$ by Rdt, where t is the time since the start of the test and R the constant strain rate of the test, the shear strain dependence in Equations 2 converts to time dependence. As in Joseph (2009, 2010), this time dependent version of Equations 2 continue to describe a dynamical system.

From Equations (2), it is possible to write $\bar{p}$ and $d\bar{p}/d\gamma$ as functions of q and obtain a single equation in terms of q as

$$\frac{d^2 q}{d\lambda^2} + C\frac{dq}{d\gamma} + Kq = 0 \qquad (3)$$

where:

$$C = [\lambda_p + J_q \exp(-\lambda_q \gamma) - J_p \tan \alpha_p \exp(-\lambda_p \gamma)]$$

and

$$K = J_q \exp(-\lambda_q \gamma)[(\lambda_p - \lambda_p) + J_p(\tan \alpha_q - \tan \alpha_p)\exp(-\lambda_p \gamma)]$$

Eq. (3) is a homogeneous, second-order, linear equation, with non-linear, non-autonomous coefficients. It has no known closed form solution. A similar approach results in a like equation for $\overline{p}$, i.e., the behavior of q or $\overline{p}$ is analogous to that of a mass-damper-spring system with both damper and spring nonlinear. Damping and spring stiffness both decay as they approach the steady-state where damping reaches a constant value and spring stiffness becomes zero.

The method to obtain the model parameters and calculate the stress-strain curves is as follows: For each test, calculate the values of $dq/d\gamma$, $d\overline{p}/d\gamma$ (and for drained tests $de/d\gamma$) from the measured data. Then, use standard fourth-order Runge-Kutta numerical integration to calibrate Equations (2) to the measured data to obtain the model parameters J, λ, and α corresponding to q and $\overline{p}$ (and for the drained case, e). Next, use these parameters in Equations (2) and starting from the reported initial conditions, calculate q and $\overline{p}$ (and for the drained case, e) to obtain the full curves for the test. Correlate this derived curve against the original, measured values. Billo (2007) details with examples the implementation and use of the Range-Kutta method in a standard spreadsheet.[2]

The next section tries to falsify the model described by Equations 2, by applying it to data from 130 undrained and drained tests on clays, sands, silts, and mixtures thereof under various conditions of OCR, density, and stress paths.

VALIDATION OF THE MODEL-UNDRAINED SHEAR

This section tries to falsify the physical model described by Equations 2 above by reanalyzing 68 undrained, monotonic shear tests from three test programs originally analyzed with the phenomenological model of the previous chapter in Joseph (2009, 2010). The three test programs were Sheahan (1991), Gens (1982), and Plant (1956)[3]. The reanalysis showed that the model matched

---

[2]In 2003, Billo had not yet written his book and so I had to create my own spreadsheet version. You can download this spreadsheet from the https://www.crcpress.com/Dynamical-Systems-Based-Soil-Mechanics/Joseph/p/book/9781138723221 for this book. It is for the same analysis, shown in Figure 3.1 of this chapter for the extension test CTx-50. Please feel free to use it as you see fit. The spreadsheet works like this: First, input the stress-strain values from a test you wish to calibrate the model to. Then put in first estimates for the parameters. Finally, go to the tools menu and choose the run command to run the Range-Kutta method programmed in and using Excel's "solver" function. I intentionally used a spreadsheet-based approach rather than some fancy program such as Matlab simply because spreadsheet programs are much more readily available and because most people are familiar with using spreadsheets.

[3]Keen students of soil shear may recognize the Plant dataset. I chose it purposely as it was the main dataset, one of two, used to inform the findings presented in the original Roscoe, Schofield and Wroth (1958) paper on the critical state. Nonetheless, even in

the data with an average coefficient of correlation of 0.95 ± 0.08 for q and 0.95 ± 0.09 for $\bar{p}$, i.e., that the fits were good and that so, the hypothesis stayed unfalsified.

Sheahan (1991) conducted 28 compression and 10 extension triaxial tests at various strain rates on uncemented, resedimented Boston Blue Clay-a glacial outwash of illitic CL clay deposited in a marine environment. Samples one dimensionally (K_o) consolidated to different overconsolidation ratios (OCRs) were sheared undrained in monotonic compression and extension at various constant strain-rates. Noticeable shear planes occurred only for a few of the tests. All the compression tests showed significant, uniform bulging at the end of the tests while all the extension tests but one exhibited necking, above 12% axial strain. Bulging in compression tests resulted in area increases in cross-sectional sample area at the mid-height of the sample by over 20% and corrected for in the data by averaging, while the necking in extension was significantly smaller, usually less than 5% of original cross-sectional sample area at sample mid-height. Sheahan did not give information as to the type (local or uniform) of the bulging.

Equations 2 fit to Sheahan's 28 compression tests with correlation coefficients that averaged 0.93 for q and 0.95 for $\bar{p}$. The corresponding correlation coefficients for the ten extension tests were 0.90 and 0.88.

Figure 3.1 shows fits of the model to compression and extension tests from Sheahan's data, chosen for the figure because their coefficients of correlation are typical for their corresponding test type. (Note that these are the same tests shown in Chapter 2, Figure 2.1 except there, the analysis was with the phenomenological model; also, as before the two tests have little in common other than that the goodness of fit of Equations 2 are typical for their test type.)

The vertical effective confining stress $\bar{\sigma}_{vc}$ normalizes the data shown. For the compression test (CTx-38), the calculated normalized peak values for $q/\bar{\sigma}_{vc}$ and $\bar{p}/\bar{\sigma}_{vc}$ of 0.53 and 1.04 compared respectively with measured values of 0.49 and 1.04 (a difference of 9.0% and 0%). The correlation coefficients of the calculated to the measured curves were 0.95 for $q/\bar{\sigma}_{vc}$ and 0.96 for $\bar{p}/\bar{\sigma}_{vc}$. For the extension test (CTx-50), the calculated normalized peak values for $q/\bar{\sigma}_{vc}$ and $\bar{p}/\bar{\sigma}_{vc}$ of -0.12 and 0.26 compared respectively with measured values of -0.13 and 0.22 (a difference of 8.3% and 18%). The correlation coefficients of the calculated to the measured curves were 0.94 for $q/\bar{\sigma}_{vc}$ and 0.97 for $\bar{p}/\bar{\sigma}_{vc}$.

---

that paper, the authors did not directly model the Plant data, as I did in this paper. I chose this data set specifically to counter a remark Andrew Schofield made to me in 2007, viz., that "... you can torture a specimen in a triaxial test to make it give you the results you are looking for." In this case, I could not have done this as Plant ran his tests a few years before I was even born. Nonetheless I strongly agree with Andrew, which is why, when I see authors confirm their model using their own test data, I smile and wonder how their papers pass peer review.

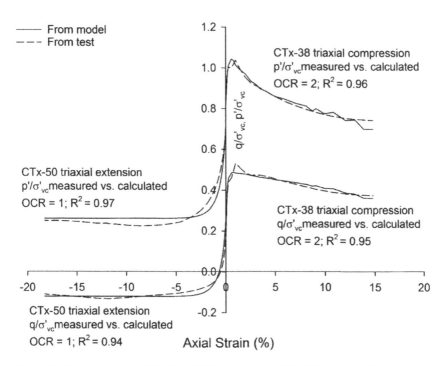

Figure 3.1 Typical measured (Sheahan, 1991) vs. calculated (Eq. 2) normalized stress-strain curves.

Figure 3.2 plots against OCR the parameter values from the compression and extension tests averaged across parameter values, regardless of strain-rate. (Note: For Figures 3.2, 3.4, and 3.5, the λ_q and λ_p values being small, they plot against a separate axis on the right.) Six out of the 168 values obtained from the 28 tests were outlier values (three λ_q and three λ_p values from tests CTx-15, CTx-40, and CTx-38) and not used in the averaging. The figure shows a smooth variation of the parameters with OCR with high correlation coefficients, i.e., following Equations 2, the parameters are strain-rate independent. As expected from the underlying stochastic model, J_q, λ_p, and α_q are very close to J_p, λ_p, and α_q respectively. The friction angles α_q, α_p increase with OCR probably due to increased particle interlocking at higher OCRs, while the equivalent surface areas to J_q, J_p and the rates λ_q, λ_p at which particles move into the flow-structure decrease with OCR.

Figure 3.3 shows the typical variation of a parameter (in this case α_q) with strain rate. Figure 3.2 shows these values averaged into the single line for α_q. Here also, no uniform variation with strain-rate is clear. The scatter in the data increases with OCR and with departure from the standard strain-rate used in triaxial tests. The analyses treat extension test strains as

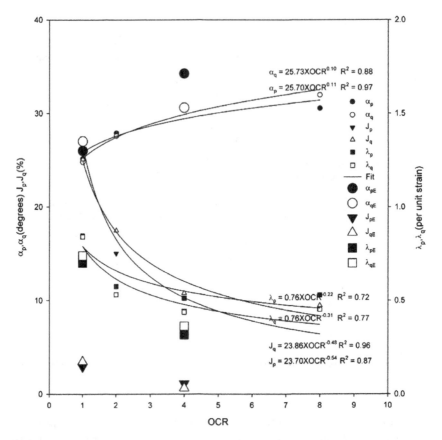

Figure 3.2 Parameter values for Sheahan (1991) compression and extension (subscript E) tests.

negative. Consequently, for Sheahan's 10 extension tests, the values of α_q and α_q are negative in order that J_q and J_p decay with strain. Likewise, for triaxial extension with effective axial stress $\overline{\sigma}_a$, and effective radial stress $\overline{\sigma}_r$, shear stress $q = (\overline{\sigma}_a - \overline{\sigma}_r)/2$ is negative and so parameters J_q, J_p and α_q, α_q are negative. However, for purposes of comparison, Figure 3.2 plots their absolute (physical) values.

Plant (1956) conducted a total of 18 consolidated undrained compression tests on – resedimented samples of London Clay-Eocene series CH clay deposited in a marine – environment-isotropically consolidated to different OCRs. Plant typically reported 17 data points per test. All samples exhibited bulging that Plant corrected by calculating sample area as the original volume divided by the axial height at any given strain. Plant provided no information regarding the formation of shear planes.

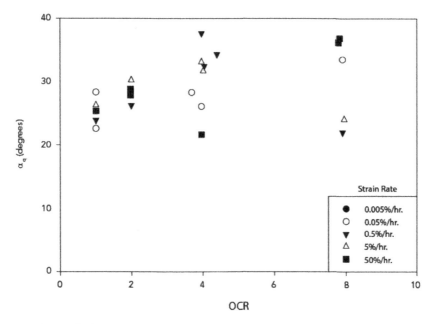

Figure 3.3 Variation of α_q with OCR and Strain Rate for Sheahan (1991) tests.

Figure 3.4 plots against OCR the average parameter values from all of Plant's tests, except for three outlier values for α_q and α_q from tests U5, U6, and U15. Equations 2 fit to Plant's 18 tests with correlation coefficients that averaged 0.98 for q and 0.95 for $\bar{p}$.

Gens conducted 15 consolidated undrained compression tests on resedimented samples of Lower Cromer Till-a low plasticity sandy CL clay-isotropically consolidated to different OCRs. Three of the fifteen tests did not appear to consolidate fully and so the model was fit only to the remaining twelve tests. Gens typically reported less than 15 data points per test. There was no information as to whether failure planes were observed or whether sample bulging occurred during the test.

Equations 2 fit to Gens's 12 tests with correlation coefficients that averaged 0.99 for q and 0.99 for $\bar{p}$. Figure 3.5 plots against OCR the average parameter values obtained from the fits.

Figures 3.4 and 3.5 plot the parameters of Equations 1 for the Gens and Plant data. As expected, model parameters J_q, λ_q, and α_q are close to J_p, λ_p, and α_p as was the case for the Sheahan data. Plant and Gens both used isotropic consolidation. Consequently, their initial configuration of particles relative to the steady-state flow-structure did not vary with OCR. Hence in their tests, J_q and J_p, both measures of the numbers of particles that need

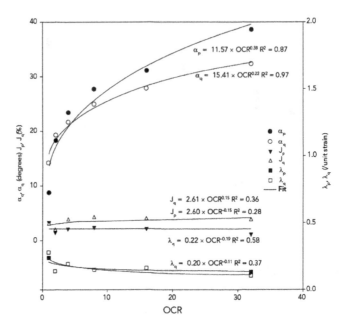

Figure 3.4 Parameter values for Plant (1956) compression tests.

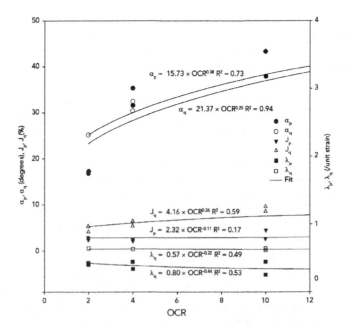

Figure 3.5 Parameter values for Gens (1982) compression tests.

to move into the flow-structure, remain approximately constant with OCR, unlike the case of the K_o consolidated Sheahan tests.

The close values of the parameters for each of the three sets of data above are no coincidence but instead a result of the 45-degree shear plane being the plane of maximum stresses as per the Mohr's circle for a triaxial compression test. The results from modeling the Sheahan data (Figure 3.2 above) show this best—the variation of the model parameters corresponding to $\bar{p}$ and q with OCR is almost identical. This even though the test samples are all different and the test data obtained over a wide range of strain rates (four orders of magnitude change). Rounding out the parameters for the equations, the six equations reduce to three, i.e., that for undrained triaxial compression, parameters J_q, λ_q, and α_q are the same as J_p, λ_p, and α_p or in other words, that $dq/d\gamma$ effectively equals $d\bar{p}/d\gamma$ for undrained triaxial compression. In other words the empirical test data show that $q = \bar{p} - \bar{p}_0 + q_0$, which is the Mohr circle for an undrained test with effective stresses on the 45-degree plane dominating the observed stress-strain-pore pressure behavior. In short, all undrained triaxial compression tests follow this simple three parameter model, i.e.,

$$\frac{dq}{d\gamma} = \frac{d\bar{p}}{d\gamma} = Je^{-\lambda\gamma}[\bar{p}\tan\alpha - q] \qquad (4)$$

and that any deviation from this is a combination of sampling/experimental error and/or statistical noise. Note that in Equation 4 above, *e* is not the void ratio, but the symbol for the Euler number, used here intentionally to show how elegant and beautiful this simple equation looks; and this simple equation describes the change in shear and confining stress on the failure plane for all undrained triaxial compression tests. This equation should also hold for extension tests, and the Sheahan test data for extension tests shown in Figure 3.2 appears to confirm this.

Recall that these parameters were from a set of stress-strain curves that went far past failure, i.e., that the model calibration values that resulted in Figure 3.2 used every single data point from the tests, including those points measured at strains well past failure, long after failure planes had developed in the sample. Recall also that the fits were very good for the entire curve, right to the end, well past the point when failure planes developed. What this means is that the DSSM equations are tracking conditions on the dominant failure plane, past failure, past the development of failure planes, and can do it very well (see the high chi values).[4]

---

[4]This contrasts with CSSM that is quite unable to track conditions once failure planes develop. In fact, the literature on CSSM cites the development of failure planes as the chief reason why CSSM based models are unable to track conditions post failure, which typically is less than 10% axial strain. As you will see in the conclusion,

Sheahan tested a lean clay, Plant a sandy lean clay and Gens a fat clay. Examining the model parameters for their data as shown in Figures 3.2, 3.4 and 3.5 respectively, it appears that both the lean clay tested by Sheahan and the fat clay tested by Gens appear to have comparable values of λ but that the addition of sand in the lean clay tested by Plant, results in λ decreasing to lower values. By and large λ appears to be insensitive to initial consolidation anisotropy as well as OCR. Likewise, α values appear in the same range for the lean and fat clay but start at a lower value for the sandy lean clay. OCR values strongly influence both α and J. This sensitivity of J to OCR appears to dominate over issues of plasticity or sand content, i.e., appears strongly controlled by soil structure. J values are similar for both Plant's sandy lean clay and Gens's fat clay showing that the presence of some sand in the clay does not seem to influence J. In short, for clays, the important variable determining λ and α is grain size while soil structure influences α and J.

VALIDATION OF THE MODEL-DRAINED SHEAR

Using the model described by Equations 2 above for the physical model, the 62 drained, monotonic shear tests from two test programs first analyzed using the phenomenological model of the previous chapter in Joseph (2009, 2010) were also reanalyzed. The reanalysis showed that the model matched the data with an average coefficient of correlation of 0.97 ± 0.03 for q, 0.93 ± 0.15 for $\bar{p}$, and 0.95 ± 0.10 for e, i.e., the fits were good and the hypothesis continued to still be unfalsified.

Arulmoli et al. (1992) tested "Nevada sand" and "Bonnie silt." The Nevada sand was a fine, uniform, angular sand that graded between U.S. No. 60 and 200 sieve sizes, had a specific gravity of 2.67, and maximum and minimum dry densities of 17.33 kN/m^3 and 13.87 kN/m^3 respectively (USCS classification SP). The Bonnie silt had a specific gravity of 2.67, and more than 80 percent of it consisted of silt-sized particles, with less than seven percent clay size. It had a Liquid Limit of 29 and a Plastic Limit of 14 (USCS classification CL). The Nevada sand triaxial tests used medium-dense samples at nominal relative densities of 40 and 60 percent, while the Bonnie silt triaxial tests were at a single nominal density of 14.5 kN/m^3.

Imre Lakatos, a noted philosopher of science coined the term "degenerate research program" for theories where excuses justify an inability of theory to match empirical data. Lakatos was commenting in general about scientific programs and likely did not even know that a field like soil mechanics existed, which makes his comments more powerful. DSSM needs no such excuses, as it can track conditions post failure plane development. CSSM on the other hand qualifies as a "degenerate theory" given its need for excuses about its inability to match the empirical data.

Triaxial tests was per USACE EM 1110-2-1906 with the Nevada sand test samples created using pluviation (a cylindrical tube with a screen at the bottom placed in a membrane stretched over a 0.0635 m diameter sample; dry sand poured into the tube and the tube and screen withdrawn using calibrated, automatic control). To create the Bonnie silt samples they created a paste of the silt at a water content of 30% that they transferred to a sample mold, then gently squeezed in the paste using a porous plunger, to reach a void ratio of 0.8.

At each density, Arulmoli *et al.* ran tests at initial confining stresses of 40, 80, and 160 kPa. They tested isotropically and anisotropically consolidated samples in compression and extension-20 tests for the Nevada sand and nine tests for the Bonnie silt. Though shear bands formed at strains of between 10 and 15 percent for all the sand tests, shearing continued to an axial strain of at least 20%. They did not report on shear plane formation for the silt samples. They also did not provide information on the end of test sample shape for either the sand or the silt samples though they report correcting sample area by dividing sample volume by sample height for each strain measurement.

The computer-controlled loadings for two of the anisotropic tests (CADE 40-96 and 60-74) on sand were non-monotonic, and this paper analyzes only the monotonic portions of the curve for these tests. Not analyzed were two tests on sand (CADC 40-109 and CADC 60-83) because they had anomalous $\bar{p}$ measurements.

For the sand, the average coefficients of correlation were 0.94 for q, 0.89 for $\bar{p}$, and 0.94 for v; for the silt, they were 0.94 for q, 0.91 for $\bar{p}$, and 0.94 for e.

Shapiro (2000) tested "Nevada II sand" and a mixture of Nevada II sand with 20% by weight of "ATC silt." The Nevada II sand was uniform fine quartz with angular to highly angular shaped particles grading between U.S. No. 50 and No. 200 sieve sizes, D50 of 0.142 mm, a specific gravity of 2.69, and maximum and minimum dry densities of 19.90 kN/m^3 and 18.62 kN/m^3 respectively (USCS classification SP). The ATC silt was quartz with a Plasticity Index of 0 and a specific gravity of 2.74. When mixed 20% by weight with the Nevada II sand, the maximum and minimum dry densities were 20.55 kN/m^3 and 18.76 kN/m^3 respectively (USCS classification SM) and with D50 = 0.125 mm. Sample preparation was here also done using pluviation, but with manual, calibrated control of withdrawal.

Shapiro conducted 12 drained triaxial tests, four on the sand and eight on the sand/silt mixture. Nine of these triaxial tests were compression tests, while three were extension tests. Tests were at a single nominal density with nominal initial confining stresses of 50, 100, and 200 kPa. One of the tests on sands (NII-20-200-CD25) had outlier values for the parameters, even though the fits were close. Shapiro did not provide any information on either end of tests sample shape or shear band formation. Additionally, Shapiro conducted

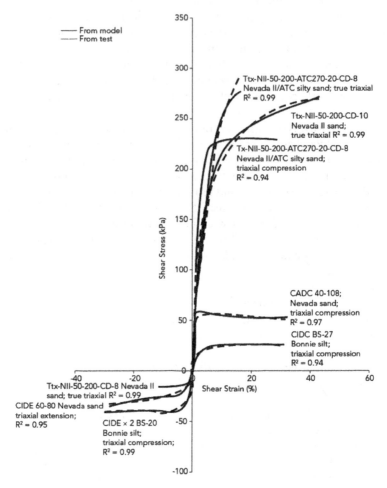

Figure 3.6 Measured versus calculated shear stress-strain curves for eight sample drained tests.

21 drained true-triaxial tests at different *b* values, five on the sand and sixteen on the sand/silt mixture. Shapiro observed the formation of shear bands for all the true triaxial tests. He thought these initiated just prior to shear failure.

The average coefficients of correlation for Shapiro's triaxial tests were 0.98 for q, 0.98 for $\bar{p}$, and 0.99 for e for the pure sand; and 0.99 for q, 0.97 for $\bar{p}$, and 0.99 for e for the sand/silt mixture. For his true triaxial tests, they were 0.98 for q, 0.98 for $\bar{p}$, and 0.86 for v for the pure sand; and 0.99 for q, 0.99 for $\bar{p}$, and 0.97 for e for the sand/silt mixture.

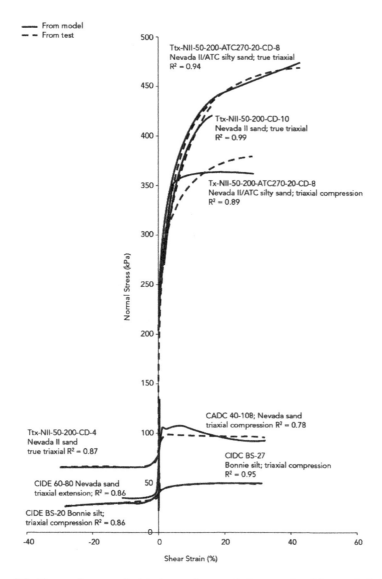

Figure 3.7 Measured versus calculated normal stress-strain curves for eight sample drained tests.

Figures 3.6 through 3.8 show typical fits of the model to data for drained compression and extension tests on the sand and silt from tests by Arulmoli *et al.* and Shapiro.

Figures 3.9 and 3.10 respectively plot parameter values for the Arulmoli *et al.* compression and extension tests on sands, Figure 3.11 for their tests

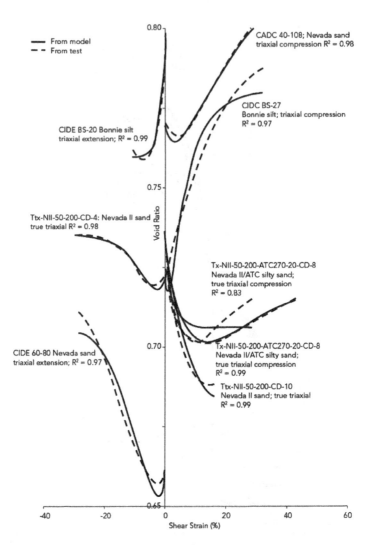

Figure 3.8 Measured versus calculated void ratio-strain curves for eight sample drained tests.

on silts, and Figures 3.12 and 3.13 respectively for Shapiro's triaxial and true triaxial tests, with the few tests for sands plotted with the tests for the sand/silt mixture.[5] The values of J_e are very small and so, plotted on a

---

[5]Steve Poulos once told me that as sand size particles exceeded 5% in a soil, then they begin to control soil behavior. This seems to be the case with Shapiro's tests with the sand/silt mixture where 20% of the mix was sand.

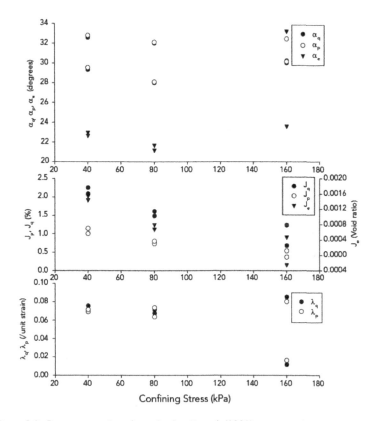

Figure 3.9 Parameter values from Arulmoli *et al.* (1992) compression tests on sand.

separate scale on the right in these figures. As expected from the underlying stochastic model, the friction angles α_q, α_p, α_e are close to each other, as also surface areas J_q, J_p, and rates of re-orientation α_q, α_p. The values of α_e show a larger variation probably due to it being calculated from very small values of void ratio changes that are difficult to measure accurately.

Arulmoli tested an angular fine uniform sand as well as a "silt" that actually appeared to be a lean clay, with 20% sand. Shapiro tested an angular to highly angular sand of similar size as that of Arulmoli's sand. He also tested this sand mixed with 20% non-plastic silt. Figures 3.9 through 3.13 show that for both the Arulmoli data and the Shapiro data, model parameters varied in an orderly manner with initial confining stress. Parameters for Arulmoli's triaxial extension tests on both the sands and the silt were roughly the same as those for triaxial compression. *J* values for the both were much lower than for clays. The high angularity sand tested by Shapiro appeared to have slightly lower α values than the less angular sand tested by Arulmoli, but otherwise parameter values seemed similar for the two

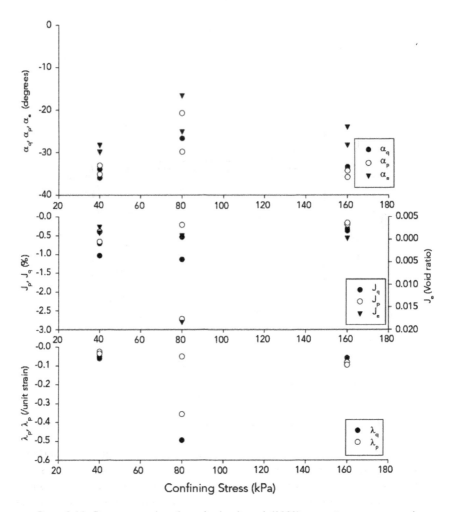

Figure 3.10 Parameter values from Arulmoli et al. (1992) extension tests on sand.

sands. Parameter values trended similarly for sand and silt, with values for the sands being slightly higher than those for the silts. In general, values decreased with increase in confining pressure.

Figure 3.13, the results of analyzing Shapiro's true triaxial tests, is particularly interesting because it shows how the model's parameters change as the stress paths rotate. When $b = 0$, test specimens are in extension and when $b = 1$, in compression. The data seem to indicate that with the exception of J_q and J_p, the other parameters remain largely unchanged. Further, J_q and J_p appear to be symmetric with b or that parameter values calculated for tests

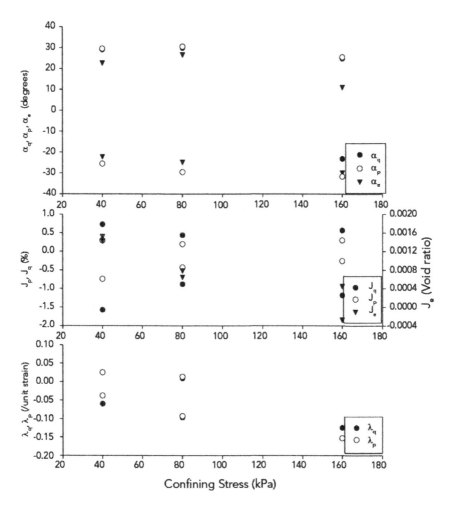

Figure 3.11 Parameter values from Arulmoli *et al.* (1992) compression and extension tests on silt.

in isotropic compression are the same as those for isotropic extension with suitable changes in sign.

PREDICTIONS

The fits to the curves shown in Figures 3.1, 3.6, 3.7, 3.8 and the high correlation coefficients reported earlier are due to fitting the model to test data, i.e. each test has its own set of (fitted) parameters different from the other

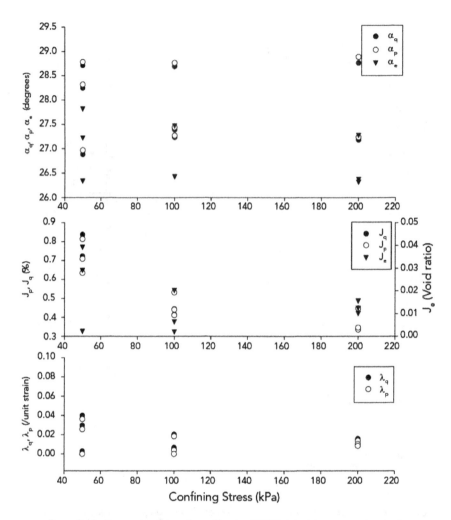

Figure 3.12 Parameter values from Shapiro (2000) triaxial compression tests.

tests. In other words, the close fits seen are not the result of an independent prediction of the test data. This section describes making an independent prediction for tests different from Sheahan's but using the model parameters obtained from Sheahan's tests and shown in Figure 3.2. The section next explores a suggestion (Castro, 2016) on how a physical feel for the model allows one to realize that parameters from one test can be used to predict behavior from certain other tests, for example, parameters from an isotropically consolidated drained triaxial compression test to predict a drained extension test.

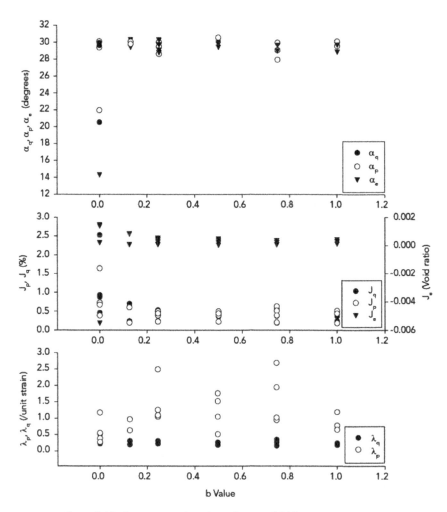

Figure 3.13 Parameter values from Shapiro (2000) true-triaxial tests.

It is very important to note that the approach taken above is "statistical" in that it calibrates the model using the average values of model parameters from multiple samples tested under nominally identical conditions. Also, the data is of a high quality, particularly, the Sheahan data. The averages reveal (as in Figure 3.2), that the parameters vary in an orderly way. Calibrating the model using single tests at each OCR or relative density would likely not reveal such an orderly variation. In short to be successful, the DSSM model needs data from multiple high quality tests run under nominally identical conditions. This is so because strength data are extremely sensitive to the initial conditions and the loading conditions through the test—slight deviations in the test result in significant differences in the measured values.

The other caveat is in the numerical analysis of the data. Great care has to be taken to ensure that one isn't finding a "local maximum" for the data fit. If the test data are of poor quality, numerous iterations may be required to find the global maximum. For tests such as triaxial undrained compression where Equation 4 should hold for a perfect test, a good place to start the iterations is with equal values for the two sets of parameters corresponding to q and p'.

It is important to obtain a physical feel for the parameters of the model as it allows for a deeper understanding for how test results vary with different stress paths, consolidation conditions, soil type, grain shape, etc. For ex., J_q and J_p the interparticle contact areas vary depending on whether the shear plane is horizontal or at 45 degrees. It also depends on the consolidation being different for isotropic consolidation where grains are likely oriented at random as opposed to say K_o consolidation where grains likely have a preferred orientation. Model parameters for the three clays tested show this clearly—for the isotropic case J_q and J_p remain largely independent of OCR as seen in Figures 3.4 and 3.5, whereas this is not the case for the K_o consolidated Sheahan test data as seen in Figure 3.2. It indicates that a soil initially in the K_o condition, is closer in terms of structure to the steady state at higher values of OCR. That this is so, is physically intuitive.

Grain shape influences initial contact areas—flat particles have different contact area than more spherically shaped ones and so is a physical determinant of the initial contact areas J_q and J_p.

Rates at which particles move to the steady-state appear to be much less sensitive to factors such as OCR as well as soil type, but the friction angle α is sensitive to OCR which causes it to vary widely for a given soil type regardless of anisotropy. More research is needed to identify these patterns based on factors such as stress paths, consolidation conditions, soil structure etc. Table 3.2 is a start at associating typical values with soil type.

Independent prediction

Fayad (1986) conducted undrained shear tests at four OCRs on resedimented, K_o consolidated samples of Boston Blue Clay. Fayad used manual stress increments to maintain K_o conditions as opposed to Sheahan who used computer controlled stress increments. Also, Fayad's reconstituted his test samples using soil used in earlier test programs and from batches different from Sheahan's. Fayad's samples had observable failure planes post failure as well as significant local bulging that he accounted for by averaging areas along the sample length.

Table 3.1 Calculated parameter values used to predict Fayad (1987) data.

OCR	I	2	4	8
α	24.8	22.2	20.9	20.1
J	23.8	16.5	11.9	8.2
λ	0.8	0.6	0.5	0.4

For these reasons, the prediction use simple averages from the curves in Figure 3.2 (and not values of specific to q and $\bar{p}$), to calibrate the model for the predictions. In other words, the predictions at any given OCR effectively use only three parameter values J, λ, and α, each being the average of their respective values specific to q and $\bar{p}$ obtained from their respective equations (shown in Figure 3.2), and corresponding to that OCR. Table 3.1 shows the parameter values calculated from the equations for the Sheahan data, at OCRs corresponding to Fayad's tests. Figure 3.14 shows the resulting predicted curves along with the actual measured values.

The fits are reasonable given that Fayad's test samples were reconstituted from previous test programs and from batches different than Sheahan's and most importantly, that Fayad used a manual procedure for K_o consolidation, resulting in some negative values in the initial measurements. The maximum fit discrepancies are at OCR 8 where for q the calculated maximum of 1.3 compares against the measured value of 1.74, a discrepancy of 25%, and for $\bar{p}$ where the calculated maximum of 1.8 compares against the measured value of 2.9, a discrepancy of 38%.

This leads to the consideration that if J, λ and α have values typical for the soil type, that this would clarify the model and, be of use in making initial first order predictions of stress-strain curves for triaxial tests on other soils. Table 3.2 is a start at providing typical values based on soil type. It reports the range of parameter values for the tests analyzed in this paper. Though the data are from a limited number of test types, it is worth noting that the range of friction angles seen for the silt and sand are comparable and the same with clays, except that expectedly, the clay explores lower friction angles. Also, reasonably, the value of λ for tests on silt and clay are comparable, while it is higher for the sands, and the value of J for the clays is higher than for the silts and sands, which in turn are comparable.

Other predictions

Castro (2016) suggested using model parameters fit to the compression test, to predict an extension test. Arulmoli's drained tests included samples consolidated isotropically to the same initial conditions and then sheared in either compression or extension. For tests CIDC_40_100 and CIDE_40_101, each

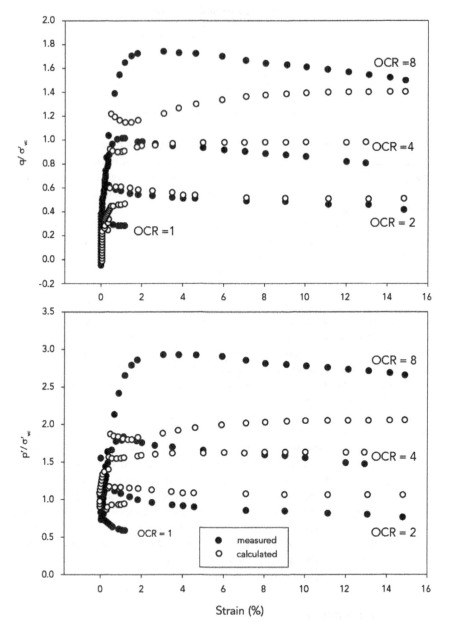

Figure 3.14 Predicted versus measured normalized stress-strain curves for Fayad (1987) triaxial compression tests.

Table 3.2 Tentative estimates of parameter values based on soil type.

Parameter	Clay	Silt	Sand
α (degrees)	15–35	20–30	20–35
λ (per unit shear strain)	0.2–0.9	0.3 to 1.0	0.5–2.0
J (% area)	3–25	0.25–2.0	0.25–2.5

sample was prepared to the same initial relative density of 40% and consolidated isotropically to a nominal pressure of 80 kPa. They were both sheared using a vertical stress path, i.e., p' kept constant while q increased, in compression for the CIDC test and in extension for the CIDE test.

For these tests with $\bar{p}$ kept constant, a little introspection shows that the Mohr Circle plane of maximum stress is again the 45-degree plane but that since $dq/d\gamma$ and $d\bar{p}/d\gamma$ are not equal, the two sets of parameters for q and $\bar{p}$, will not be the same. Also, since p' is nominally constant, a good place to start the modeling is with the equation for $d\bar{p}/d\gamma$ in the model set to zero, approximating λ_q and λ_p to be equal and solving for α_p, then setting the initial value in the Range-Katta model for both α_q and α_p to the calculated α_q value. In reality, as described below, it is not possible for p' to remain exactly constant throughout the test, i.e., the actual plane of maximum stress in the real test, must be slightly different than the 45-degree plane, and also, changing continuously. Figure 3.15 shows the model fit to the CIDC test and the prediction of the CIDE test. The prediction uses the actual average compression stress applied during the extension test, a value of about 69 kPa, and not the nominal 80 kPa intended.

Studying this figure, it appears that the model over-predicted the shear stress results from the extension test. Closer scrutiny of the data show that correction for test specimen bulging for the compression test was per USACE EM 1110-2-1906, an approximate correction obtained using the formula: $A_c = A_o/(2 - \varepsilon_v)$ where A_c is the area corrected for bulging, A_o the original area and ε_v is the vertical strain. This correction averages the diameter across the entire height of the sample, and uses this averaged area in the calculation stresses. USACE EM 1110-2-1906 is a standard for a triaxial compression test and does not specify a triaxial extension test. However, assuming Arulmoli et al. used this same correction with a reversal of sign for the extension tests, then disparities in area should not be a factor unless of course, deviation occurred from a circular cross-sectional shape (Castro, 2016).

Additionally, the approach used compared two single tests directly against each other, i.e., the approach was not statistical. Further, the test used a challenging stress path where confining stress was to be constant throughout the test. A close examination of the confining stress-strain curve shows that

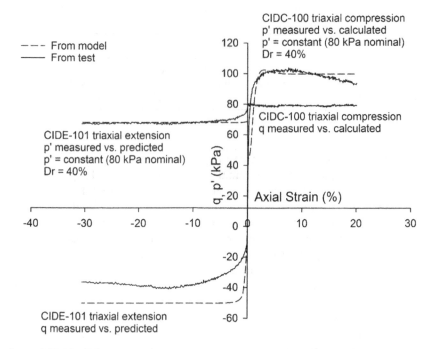

Figure 3.15 Triaxial extension test stress strain curves predicted from triaxial compression test data from Arulmoli *et al.* (1992).

this was only approximately the case. In fact, the equations of the model show why keeping the confining stress constant can only be an approximate target. When shear stress changes, particle movement occurs and this in turn causes changes in the confining stress. Hence, the computerized tests system must immediately adjust the stresses so that the confining stress remains unchanged. It can do this only after sensing a change in the confining stress, i.e., the confining stress constantly fluctuate about a mean, and is only approximately constant. This is the reason a statistical approach is far preferred when using the model to make predictions.

In general, only if the initial sample states are isotroic, and the applied stress-paths the same, or as in the case above, "mirror-symmetric," can parameters from one test be applied to another test. In the tests shown in Figure 3.15, this was the case—the samples were both isotropically consolidate and the applied stress paths were "mirror-symmetric." This is not the case though for K_o consolidated samples. Here, though the initial conditions are in an absolute sense the same in a given sample, in a sense relative to the steady-state structure, they are not the same initial condition for a compression test, as for an extension test. This is because the particles move differently in these tests. In other words, to have the same initial condition relative to the steady-state regardless of compression or extension means that samples must be isotropically consolidated to begin with. Additionally, the Mohr Circle plane of maximum stress must be the same during the tests, as also, the resultant force on the particles on average be in the same direction (or mirror symmetric).

To understand this better, examine the model (Equations 2). For a given initial condition, and for given model parameters, there is only one possible value for both for a given increase in shear strain. This means that starting from the same initial condition, model parameters from one stress-path will always result in the same stress-strain curve, i.e., given that the stress-strain curves from another stress-path will be different, model parameters for a given stress path cannot apply to any other stress path. The exception is for mirror symmetric stress paths, where with sign changes, the model parameters from the original test when used in the "mirror-symmetric" test give a first order prediction of the stress-strain curves. The extension test CIDE_40_101 has a stress path that is "mirror-symmetric" with CIDC_40_100, and so, it is reasonable to use model parameters obtained by fitting CIDC_40_100 to obtain a first order prediction of the curves that would result from the CIDE_40_101 test.

CONCLUSION

This chapter described a stochastic, micro-structural model based on the simple principle that particles move into the steady-state structure at random shear strains. It extends the phenomenological model of the previous chapter by giving it its physical basis. The resulting model's value lies in its ability to closely fit, with orderly varying parameters, a range of data from 130 drained and undrained triaxial and true-triaxial shear tests with various stress-paths (including compression and extension), on sands, silts, and clays at various densities and OCRs, and exhibiting strain-hardening or strain-softening behavior.

Present limitations of the model are that it only applies to static loading. Future research should study how it applies to unloading, dynamic loading, as well as how parameters vary with test type and soil type. The physical basis of the model parameters offers an opportunity to study how the physical properties of soil particles influence these parameters. The implications of soil shear behaving as a dynamical system also merit study.

The next chapter discusses implications of soil shear being a dynamical system by, examining the empirical evidence available from the last century or so of research on soils. Before reading this next chapter though, visualize a Poisson process–start with a radioactive material and imagine radioactive particles leaving it at random and the remaining number of particles decaying exponentially with time. Now repeat this exercise, but for soils with load carrying particles moving out of their initial positions to the steady-state position so that the number of load carrying particles that are not in their final position, decay exponentially with shear strain. Like before, write the equations for the model (Equations 2) without looking, nor from memory, but instead, by drawing from your understanding of the physical processes that brings them about.

Now here is a "thought experiment" (Castro, 2016): imagine samples of sand compacted to the same relative density, but with different tamping energy used to prepare the sample (As part of his PhD under Arthur Casagrande at Harvard, Dr. Gonzalo Castro prepared such samples). Now imagine shearing these samples and fitting the model to the data. How would model parameters to vary with tamping effort? Next think about one-dimensional consolidation. Can parameters from some special triaxial test predict the one-dimensional stress-vertical strain curve?

Success in exercises like these implies a thorough understanding of DSSM. It is the foundation needed to fully understand the remaining chapters.

Chapter 4

Soil shear–the physical model (part 2)

This chapter delves deeper into the implications of the physical model. As always, proceed with this chapter only when able to visualize the equations of the model in Chapter 3 and write them out without looking, based not on rote memory, but from a deep understanding of how they came about.

As per the phenomenological model of Chapter 2, i) the rates of change of shear stress, effective normal stress, and void-ratio are proportional to the applied values of shear and confining stress, with the initial proportionality values decaying exponentially with shear strain to reach ultimately a value of zero at the steady-state condition and ii) shear stress works to destroy soil structure, while confining stress works to preserve it. Chapter 3 described the underlying physical basis for this phenomenological model. This physical basis derived from two scientific hypotheses: i) simple friction acting at interparticle contact areas of the load carrying particles governs soil shear; ii) particles move into the steady-state flow structure at random shear strains, i.e., as a Poisson process, causing this interparticle contact area over which the friction acts to decrease exponentially with strain. This results in the model described in the previous chapter and specified by Equations 2 of that chapter.

The model is implicitly time dependent—to express it explicitly in terms of time t use the relationship $dq/d\gamma = dq/dt * dt/d\gamma = dq/dt * 1/R$, where R is the instantaneous strain-rate $d\gamma/dt$. The parameters of the model are themselves strain-rate independent, except for the secant friction angles $\alpha_q, \alpha_p, \alpha_e$—to the extent that these friction angles depend on the material's underlying coefficients of static and dynamic friction, strain rate plays a role in the behavior. In the context of a constant strain-rate shear test, replacing shear strain γ by Rt and the change in shear strain $d\gamma$ by Rdt, where t is the time since the start of the test and R the constant strain-rate of the test, both strain rate and time become explicit.

Note that the right-hand side of the model equations is in terms of strain, not strain increment. In general, an element of ground that has been perfectly sampled or perfectly tested in-situ is at an unknown initial strain. Hence, strain increment is the preferred term with which to formulate constitutive models because strain increments are knowable, being a measurable change in element geometry. Alternate "state variables" may suffice, void-ratio for example, for this purpose—systematically replace the shear strain in the equations for the model to express the model in this form. At the steady-state the model's equations will continue to evaluate to zero because the interparticle contact areas of particles in the shear zone that are not at the steady-state are all. Chapter 6 demonstrates this change of the model to using void-ratio, zero, because there are no such particles not in the steady-state flow structure. When specializing the model for the case of one-dimensional consolidation.

As in Joseph (2009, 2010), the Poisson model derived in Chapter 3 continue to describe a dynamical system with a fixed, single-point attractor (the steady-state condition) defined by the steady-state void-ratio (and negligibly, the strain-rate). Given that the attractor is a fixed point, behavior is not chaotic; stress-paths for various initial conditions converge to this steady-state point attractor as described in Poulos (1981). A measure of convergence used in dynamical systems theory is the Lyapunov exponent. For soils the Lyapunov exponents are λ_q and λ_p and are both negative. They must be because negative Lyapunov exponents are characteristic of dissipative systems such as is the case for soil shear. Thermodynamically speaking, dissipative systems are non-conservative systems. A dissipative structure is a dissipative system that has a reproducible steady-state. Natural evolution of the system or artificial means, or a combination of the two can drive the system to this steady-state. Particulate materials such as soils are dissipative structures.

The global phase space of the soil deformation dynamical system has three dimensions, q, $\bar{p}$ and e. The global-attractor is the steady-state line in positive q, $\bar{p}$, e space (e will not physically be able to take on negative values despite the form of Equation 1c). All the steady-state attractor points make up the steady-state line. Poulos *et al.* (1985) suggest as a heuristic for clayey materials that the projection of the steady-state line onto the e, $\bar{p}$ plane is parallel to the materials normal consolidation line. The steady-state line will be slightly different to the extent that each steady-state point that makes up the line depends slightly on strain-rate. In other words, if taking strain-rate into account, the steady-state line is a two-dimensional surface in a 4-dimensional phase space, the new dimension being strain-rate; it is a one-dimensional line in a 3-dimensional phase space otherwise (no strain-rate dimension).

SOIL PARTICLES MOVE AT RANDOM

Not explained in the Chapter 3 were the reasons why i) particles move into the steady state position at random shear strains and ii) why the total interparticle contact area decreases as the number of particles. As described in Joseph (2013a), the reason for both these behaviors is that particles in the shear zone that carry the load, move at random during shear. Physical considerations suggest that for particles to move into the steady-state flow-structure at random shear strains, they must first move randomly. In other words, the random movement of load carrying particles in the shear zone causes particles to move into the steady state position at random shear strains. More rigorously, if load-carrying particles in the shear zone move randomly when the loads they carry changes, then this means that any given particle will move into the flow-structure at some random shear strain. This in turn means that the shear strain for which the load-carrying particle is not in the flow-structure must be a continuous, positive random variable X. That movement into the steady-state flow structure occurs at some random X, in turn requires that the probability of this event occurring at any shear strain is the same i.e. that:

$$\text{Prob}\{\chi > \gamma + h | \chi > \gamma\} = \text{Prob}\{\chi > h\} \tag{1}$$

where shear strain h occurs after shear strain γ.

This is exactly analogous to the case of radioactive decay where radioactive particles leave the radioactive material at random times. As detailed in Appendix 1, for such processes, particles are not in the flow-structure with a probability $e^{-\lambda\gamma}$, and in the flow-structure with a probability $1 - e^{-\lambda\gamma}$, where λ is the rate (per unit strain) at which particles move into the flow-structure. This results in a binomial distribution where for an initial N_0 particles, independent, and identically distributed in X, the probability of N_γ particles not being in the flow-structure at shear strain γ is:

$$\text{Prob}\{N_\gamma = n\} = \frac{N_0!}{n!(N_0 - n)!} \exp(-n\lambda\gamma)[1 - \exp(-\lambda\gamma)]^{N_0 - n} \tag{2}$$

where N_γ is the number of load carrying particles in the shear zone that are not in the steady-state position at shear strain γ. Therefore, the expected value of N_γ is:

$$E[N_\gamma] = N_0 \exp(-\lambda\gamma) \tag{3}$$

As described in the previous chapter and in Joseph (2012), as the number of particles N_γ decays exponentially, so also does the total interparticle contact area of the particles that constitute N_γ. This implies that on average, the interparticle contact area between any two individual particles not in the steady-state position is always the same as shear proceeds and that consequently, total interparticle area of the particles not in the steady-state decreases in step with the number of particles that are not in the steady-state, i.e., also decays exponentially. Note that once particles move to their steady-state position, the flow-structure, a statistically constant structure, starts to develop, i.e., statistically speaking, once a particle moves to its steady-state position, it effectively remains in that position for the remaining deformation, unless there is a discontinuous change in the stress-path.

The exponential decay of particles not in the steady-stat can happen only if the particles are load carrying particles that move into the steady-state at random, i.e., effectively, in proportion to the number of particles corresponding to any given interparticle contact area, during the shear process. This in turn means that they maintain the same average interparticle contact area for the distribution as a whole. This movement to the steady-state can happen in this uniform manner across the entire distribution of interparticle contact areas only if the probability of it happening at any given value of interparticle contact area is the same for any interparticle contact area. This in turn implies that this probability is the same for any given particle, i.e., that movement to the steady-state position of any given particle occurs at a random shear strain, as previously defined by the Poisson process driving the model.

More rigorously, at any given strain, the percent number of load carrying particles not in the steady-state position has some distribution with respect to the interparticle contact area of these individual particles. Let this distribution be as follows:

$$P_n = f(a) \tag{4}$$

where: P_n is the percent number of particles that have interparticle contact area a. Let the expected value $E[a]$ of this distribution be a_{avg}, the expected value of the distribution P_n defined in Equation (4). The assumption in Joseph (2012) is that a_{avg} stays unchanged through the shear process. This can only happen if the distribution defined by Equation (4) still is unchanged with shear, i.e., that P_n still be the same throughout the shear process. This can only be possible if the number of load carrying particles moving into the steady-state is in the same proportion as the original number of load carrying particles not in the steady-state corresponding to any given interparticle contact area. This is exactly what would occur if, load carrying particles are moving into the steady-state at random shear strains, as per a Poisson process.

To help clarify this, consider the highly artificial example of 100 load carrying particles in the shear zone that are not in the steady-state position and with only 2 interparticle contact areas a_1 and a_2 with 90% of these particles with interparticle contact area a_1 and the remaining 10% with interparticle contact area a_2. With shear, for the average interparticle contact area to still be the same, the percent distribution P_n must still be the same. This requires that at any given strain, the numbers of load carrying particles corresponding to a_1 and a_2 that move to the steady state be in the same proportion as their original numbers.

As an example, at some arbitrary strain, assume that 50 of these particles have moved to the steady-state. For the remaining 50% particles to have the same distribution P_n, it requires that the percent number of particles corresponding to a_1 and a_2 still be unchanged at 90% and 10% respectively or at 45 and 5 respectively. Correspondingly, this requires that the number of particles that did move into the steady-state during the shear till this strain were also 45 and 5, i.e., each exactly in the same proportion (50%) as their original values. In other words, the average interparticle contact area across the entire distribution is the same all through shear deformation. This can only occur if the movement is happening uniformly for all particles at any given interparticle contact area a. This uniformity in turn can occur only if each particle has the exact same probability of moving into the steady-state, i.e., behaves as per Equation (1).

The specific type of distribution i.e., the specific form P_n has does not matter for purposes of the analysis. As particles move into the steady-state position from randomly within this distribution, because of the randomness, statistically, equal numbers of particles for each interparticle contact area will move into the steady-state flow-structure with the result that the distribution itself—the percent number of particles not in the steady-state position with respect to the interparticle contact area—remains unchanged with shear. Since the distribution stays unchanged with shear, so also the expected value of the interparticle contact area between individual particles stays unchanged. Consequently, though the number of load carrying particles in the shear zone decreases with shear, the mean value of interparticle contact area between individual load carrying particles stays unchanged with shear (equal to the expected value of the distribution), while at the same time, the total absolute interparticle contact area between all the load carrying particles decreases.

Before the start of shear let the shear zone contain N_q and N_p independent, identically distributed load carrying particles per unit volume of the sample that are not in the steady-state flow structure and whose movement into the steady-state flow-structure affects q and $\bar{p}$ respectively. Chapter 3 showed that as shear continues these particles move into the steady-state position at rates λ_q and λ_p per unit strain that for any given shear strain γ the expected value of the number of particles that effect q and which are not in the flow structure are $N_q e^{-\lambda_q \gamma}$ and $N_q e^{-\lambda_p \gamma}$ corresponding to the shear and normal stresses q and $\bar{p}$. Likewise, the expected value of the number of particles that effect $\bar{p}$ which are not in the flow-structure are $N_p e^{-\lambda_q \gamma}$ and $N_p e^{-\lambda_p \gamma}$ corresponding to the shear and normal stresses q and $\bar{p}$.

As described above, each load carrying particle has the same expected value for interparticle contact area—that of the distribution of which it is a part. Consequently, the total interparticle area corresponding to N_q and N_p particles decreases as N_q and N_p. In other words, if the initial total interparticle contact areas corresponding to N_q and N_p load carrying

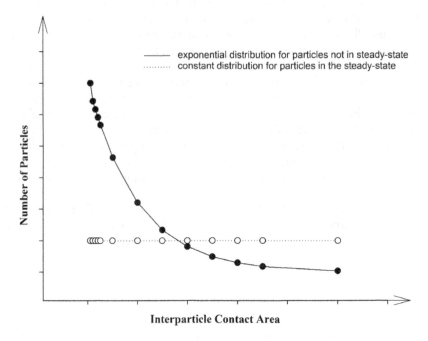

Figure 4.1   Schematic of number of load carrying particles vs. interparticle contact area.

particles in the shear zone be J_q and J_p respectively, then these interparticle contact areas at any given shear strain γ that affects q are $J_q e^{-\lambda_q \gamma}$ and $J_q e^{-\lambda_p \gamma}$ and that affects $\bar{p}$ are $J_p e^{-\lambda_q \gamma}$ and $J_p e^{-\lambda_p \gamma}$. From this follows the model proposed in Chapter 3.

Though the specific distribution of the percent number of load carrying particles not in the steady-state position with respect to the interparticle contact area does not affect the analysis presented above, the Maximum Entropy Production Principle indicates what this distribution should be. The Maximum Entropy Production Principle states that a system tries to maximize its entropy at all times. An important result that follows from this is by Jaynes (1957), who showed that to maximize entropy, any system that is varying such that the mean is positive and constant, must follow an exponential distribution—see for example Gokhale (1975) for the detailed derivation. This means that the number of load carrying particles decays exponentially as their corresponding interparticle contact area increases. Figure 4.1 schematically shows this distribution.

Particles in the steady-state no longer vary in their distribution and in this case, again following from Jaynes (1957), Gokhale (1975), to maximize entropy, the percent number of particles varies uniformly with interparticle contact area as also shown schematically in Figure 4.1. Physical considerations show that at the steady-state flow-structure, this must indeed be the case.

To summarize, load-carrying particles in the shear zone move at random. This in turn implies that 1) load carrying particles in the shear zone move into the steady-state position at random shear strains and 2) the expected value of the interparticle contact area between load carrying particles that are not in the steady-state position remains unchanged with shear. As was described in Chapter 3, the consequence of particles in the shear zone moving into the steady state at random shear strains is that the number of load carrying particles in the shear zone that are not in the steady-state position decays exponentially with strain. The consequence of the expected value of the interparticle contact area between any two particles in the distribution of load carrying particles in the shear zone not in the steady-state position staying constant as shear proceeds is that the total interparticle contact area of these particles decays as the number of these particles. This random movement of these load carrying particles during shear is the underlying basis of the dynamical systems model that closely fit, with orderly varying model parameters, a range of shear tests from 130 drained and undrained tests on sands, silts, clays at various densities and OCRs, along various different stress-paths, and with strain-hardening or strain-softening behavior.

MODELLING BEHAVIOR POST SHEAR LOCALIZATION

A feature of dynamical systems models is that in theory, accurate information of the system from any single point is sufficient for the model to generate the system's entire behavior accurately. In practice, however, measured data points are not completely accurate due to experimental errors and use is made of available measurements over and above the minimum needed in order to i) reduce the effect of these errors and ii) to help with the calibrating the model's parameter values. This feature of dynamical systems to be able to work with a partial data set allows the model described in Equation 1 to predict shear behavior post localization.

Post shear localization, measured values of stresses and void ratios are not accurate for either drained or undrained tests. This is because the specimen is usually no longer a right cylinder, void ratios in the localized shear zones are different from in the rest of the sample, and the system—the localized shear zone and the surrounding material—governs observed behavior. Once localized deformation develops, models that determine their parameters based on the assumption of homogeneity of deformation for the entire stress-strain curve are invalid. In contrast, the dynamical systems based soil model does not make such an assumption of homogeneity of deformation.

Additionally, in the pre-peak region of the test data, the rates of change for q and $\bar{p}$ are high as compared with the post-localization region of the curve. Consequently, tests have more measurements made in the pre-peak region, with changes in strain between measurements usually being far less than one. After the development of failure planes, the curves tend to be flat, with small changes in q and $\bar{p}$, and with measurements made at larger intervals of strain. Therefore, the high rates of change of q and $\bar{p}$ in the pre-localization part of the curve dominate model calibration and largely define the model's parameters, with the result that the model is insensitive to post-localization measurement values. This is the falsifiable hypothesis that this chapter tests with experimental data, i.e., this chapter uses the standard method of science—testing hypothesis with empirical evidence (the test data), in an attempt to falsify the hypothesis.

Application of the model to special tests run by Castro (1969) does not falsify this hypothesis and confirms the model's ability of Equations 2 to predict post localization behavior. Castro explored the influence of the failure planes on test measurements. He ran two tests on sand commercially available under the trade name "Banding Sand", manufactured from St. Peter Sandstone. The sand consisted of uniform, clean, sub-rounded to sub-angular fine quartz grains with specific gravity 2.65, D_{10} size 0.97 mm, and coefficient of uniformity 1.8 (USCS classification SP). Its friction angle ranged from 30 degrees for the loose state to 40 degrees for the very dense state, with a friction angle of 30 degrees at the steady-state.

Castro used special end platens to ensure uniform stress conditions through the sample. One test (Test 1-4) had a sample length (L) to diameter (D) ratio of two, while the other (Test 1-5) had a L/D ratio of one. Both samples had the same initial stress history and the same initial isotropic confining stress of 100 kPa. Failure planes occur for tests on dense soils whose initial void ratio is well below the steady-state line, and, consequently, Castro prepared all his samples with the same initial void ratio of 0.52, which was well below the steady-state line void ratio of approximately 0.72, corresponding to the 100 kPa confining stress. Castro built the samples by tamping layers of soil, with tamping controlled based on earlier tamping experiments, to achieve the desired void-ratio.

On drained shear, as expected, both samples showed strain-softening behavior together with well-developed failure planes. For test 1-4 with the L/D ratio of two, the failure plane was contained entirely within the sample and the measured stress-strain and especially, void ratio curves showed a sharp break at the point that the shear plane developed and beyond which, measurements of the void ratio were in error.

For test 1-5 with the L/D ratio of one, unlike test 1-4 with L/D ratio of two, the sample did not contain the entire failure plane. Rather, multiple failure planes cut through the entire sample from end-platen to end-platen. The measured stress-strain curves had no sharp break, and the measured void ratio smoothly headed to its correct and expected steady-state value at the end of the test.

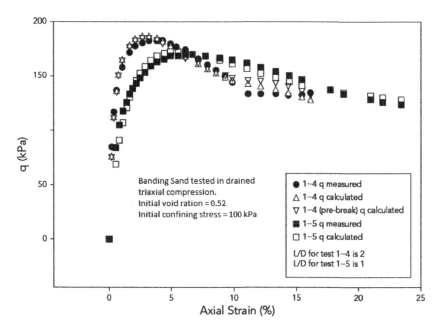

Figure 4.2 Shear stress-strain curves with/without failure plane (data from Castro, 1969).

Figures 4.2 through 4.4 show the results of fitting the model to the entire curves for both tests, as well as to only the part of the curves before where the break occurred for the test with L/D = 2. Table 4.1 shows the corresponding model parameter values.

The analysis shows that the value of about 34 degrees for α_q and α_p is robust for all three model fits. This value compares with the average of the friction angle values reported by Castro for this sand. The values of λ_q and λ_p are approximately 0.1 for the test with no break and are the same for the fit to the pre-break portion of the curves with the break. Including the break, this value goes to about 0.18, i.e., the decay rate increases in order that the model can fit the lower values post-break.

Of interest is the difference in interparticle contact area J between the tests with L/D of one and two. The failure plane for the test with L/D of 1 intersected the frictionless end platens at both ends and so these intersected frictionless surfaces of the end platens effectively form part of the failure surface. The values J_q and J_p stand for the number of particles per unit area of the failure surface that offer frictional resistance but which are yet to move to the steady-state. The number of these particles is not significantly different for the two tests. However, because the test with the L/D ratio of 1 includes area intersected by the failure plane with the frictionless end platens

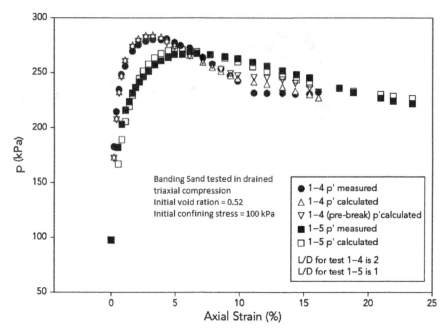

Figure 4.3 Normal stress-strain curves with/without failure plane (data from Castro, 1969).

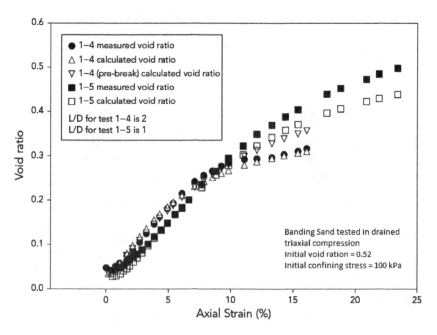

Figure 4.4 Void ratio-strain curves with/without failure plane (data from Castro, 1969).

Table 4.1 Parameters from fitting the model to Castro (1969) tests with (test 1-4, L/D = 2) and without (test 1-5, L/D = 1) breaks in curves because of shear plane formation. All samples were at an initial void ratio of 0.52.

Parameter	Test 1-5 L/D = 1 (no break)	Test 1-4 L/D = 2 (only data pre-break fitted)	Test 1-4 L/D = 2 (full curve analyzed)
α_q (degrees)	33.5	33.9	34.1
α_p (degrees)	33.5	34.0	34.1
λ_q (alignment per unit shear strain)	0.09	0.11	0.17
λ_p (alignment per unit shear strain)	0.08	0.12	0.18
J_q (% area)	4.97	10.3	10.2
J_p (% area)	5.00	10.2	10.0
R_q^2	0.97	0.99	0.98
R_p^2	0.97	0.99	0.98
R_e^2	0.99	0.99	0.99

at both ends, this value effectively distributes over a larger failure surface area. Consequently, the values of J_q and J_p for the test with L/D of one are lower as compared with the test where L/D is two.

In general, the break in the curve does not influence peak values, and the model fits the data well in the early part of the curve regardless of whether there is a break post-peak. The model also predicts behavior post localization well.

PHYSICAL BASIS OF NORMALIZED BEHAVIOR

Since the 1950's it has been known that for most clays, the undrained shear-stress vs. strain and normal stress-versus strain curves plot almost identically when normalized by the original effective confining stress. One version of this relationship, used extensively by practicing engineers, is the empirical observation that the ratio of the undrained shear strength to the original consolidation stress is a constant for a given OCR. Slope stability and finite-element analyses of geotechnical engineering problems make extensive use of this relationship.

It appears that Skempton was the first to formalize this constant ratio for the case of normally consolidated clay, noting that for most clays this ratio was around 0.3 (Skempton, 1948; Niechcial, 2002). Also, well known since the 1950's mostly from work by Henkel in Imperial College (see for example, Henkel 1956, 1960 and Henkel and Wade, 1966), is that stress-strain curves from the undrained shear of clay at any given OCR empirically demonstrate normalizability. In other words, normalizing the entire shear-stress vs. strain or confining-stress vs. strain curve by the initial consolidation stress, results in a single curve for each, corresponding to a given OCR. A consequence of this is the constant normalized peak strength $S_u/\overline{p}_c$ for a given OCR. As this chapter will show, this behavior is due to simple friction at interparticle contacts, governing soil shear.

One physical explanation for this has been the general statement that stress-strain curves should normalize if shear behavior is frictional. However, no detailed account of the precise frictional mechanism that would account for this observed behavior appears in the available published literature. This section (originally described in Joseph 2013b) uses DSSM theory to show why stress-strain curves normalize for undrained shear. DSSM is the only theory to date that does not simply take normalized behavior as a given but instead, can offer a detailed physical account of normalization behavior from a fundamental, microstructural basis.

Consider the prototypical case of simple friction—a block sliding on the surface of a table and with some coefficient of friction between the materials of the block and the table surface. If the normal force is some multiple of the block's weight, then the shear force is correspondingly the same multiple of the shear stress for the condition of sliding with only the brick weight being the normal force. More precisely, the ratio of the shear to the normal force is the dynamic coefficient of friction. In short, for any situation where the operant mechanism is simple friction, the shear stress, normalizes with respect to the applied normal stress.

For the case of soil shear, the change in stresses from their initial values is a function of the decrease in the interparticle contact area of particles that are not yet at the steady state. Initially, this area has some value, and with strain, this area decreases exponentially. From Figures 3.2, 3.4, and 3.5 of Chapter 3, the parameters that govern are essentially functions of OCR. In other words, for samples at a given OCR, and at a given strain, the values of these constants are the same for all such samples, for shear along the same stress-path.

All such samples at a given OCR, strain, and on the same stress-path are analogous to the case of the same brick sliding on the surface of a table. Samples at the same OCR start initially with the same interparticle contact area (for a given stress path), and this contact area decreases identically with any other sample likewise prepared at the same OCR, regardless of the initial confining stress, and sheared along the same stress path.

To be specific, from Figure 3.2 of Chapter 3, for the case of Boston Blue clay, as expected, the values of model parameters J_q, λ_q, and α_q are close to J_p, λ_p, and α_p. Let us for convenience consider them the same and equal to J, λ, and α respectively for some particular OCR. Hence, for all samples at this OCR, at any given strain, all samples will have the same values of J, and α operant at that strain. Likewise, from the start of shear, until reaching the steady-state, the values of these parameters for a sample at the same OCR and on the same stress path are identical for each corresponding value of strain.

In other words, at any given strain, each sample at the same OCR is analogous to the same block on the same table, with the same frictional parameters acting on it. Consequently, given that this applies to the entire stress-strain curve from the initial condition to the steady-state condition, the changes in stresses normalize, as was the case for the block on the table.

In the case of isotropic consolidation, the initial shear stress is the same, i.e., zero stress. Likewise for K_o consolidated samples, the initial shear stress is $\frac{\sqrt{2}\bar{\sigma}_{vc}(1-K_0)}{3}$ which after overconsolidation becomes $\frac{\sqrt{2}\bar{\sigma}_{vc}(1-K_{ocr})}{3}$ where K_{ocr} corresponds to the coefficient of lateral earth pressure at rest for the over consolidated sample, and is the same for a given OCR. In other words, initial shear stress normalizes with respect to $\bar{\sigma}_{vc}$ as $\frac{\sqrt{2}(1-K_{ocr})}{3}$ which is the same for samples at the same OCR, regardless of the particular value of the consolidation stress. Subsequent changes in shear stress at any given strain likewise normalize due to the operant mechanism of simple friction. Consequently, the entire shear-stress versus strain curve likewise normalizes. Since the shear-stress versus strain curve normalizes, so also must the peak value S_u and hence the ratio $S_v/\bar{p}_c$ is a constant for a given OCR.

For the same reasons, the normal stress curve likewise normalizes—the same conditions hold—the initial vertical stress always normalizes to 1. Thereafter, changes normalize also, as per the mechanism of simple friction operating throughout the shear deformation to the steady-state.

STRUCTURAL BASIS FOR $\ln(S_v/\bar{p}_c)$ VARYING LOGARITHMICALLY WITH OCR

In Chapter 3, Figures 3.2, 3.4, and 3.5 show how the parameters values vary with OCR. However, these curves are from individual tests at different OCRs, and in Figure 3.2, the tests used were at various strain rates. While the theory used to determine the model parameters says that none of this should matter, it would be prudent to check that these curves indeed predict overall soil behavior. One way to do this is to use them to predict well known, general soil properties that are valid across any OCR, such as the variation of $\ln(OCR)$ with $\ln(S_u/\bar{p}_c)$. This section investigates how closely these curves predict this variation.

A measure K of the non-dimensional interparticle contact area of the particles not in the flow-structure at any given strain is the summation of the coefficients of Equations (2) in Chapter 3 integrated between the limits $\gamma = \infty$ (where in the flow-structure, $K = 0$) and any given strain $\gamma = \gamma_i$ as:

$$
\begin{aligned}
K_q &= \int_{\infty}^{\gamma_i} J_q \exp(-\lambda_q \gamma)d\gamma + \int_{\infty}^{\gamma_i} J_q \tan \alpha_q \exp(-\lambda_p \gamma)d\gamma \\
&= \frac{J_q}{\lambda_q} \exp(-\lambda_p \gamma) + \frac{J_q \tan \alpha_q}{\lambda_p} \exp(-\lambda_p \gamma)
\end{aligned}
\tag{5a}
$$

$$K_p = \int_{\infty}^{\gamma_i} J_p \exp(-\lambda_q \gamma) d\gamma + \int_{\infty}^{\gamma_i} J_p \tan \alpha_p \exp(-\lambda_p \gamma) d\gamma$$

$$= \frac{J_p}{\lambda_q} \exp(-\lambda_q \gamma) + \frac{J_p \tan \alpha_p}{\lambda_p} \exp(-\lambda_p \gamma) \tag{5b}$$

where K_q, K_p are the non-dimensional interparticle contact areas corresponding to q, $\bar{p}$ respectively of particles that are not in the flow-structure at shear strain γ_i. Equation (5a) is the non-dimensional interparticle contact area of particles not in the flow-structure that corresponds to the shear stress observed at shear strain γ_i. Hence at the strain corresponding to the peak shear strength S_u, equation (5a) gives K_{S_u}, the non-dimensional interparticle contact area of particles not in the flow-structure at the strain corresponding to the peak shear strength S_u. Figure 4.5 plots $\ln(K_{S_u})$ versus $\ln(\text{OCR})$ and $\ln(S_u/\bar{p}_c)$ for the Sheahan (1991) data. The linear relationships between $\ln(K_{S_u})$ versus $\ln(\text{OCR})$ and $\ln(S_u/\bar{p}_c)$ imply that a similar linear relationship exists between $\ln(\text{OCR})$ and $\ln(S_u/\bar{p}_c)$, i.e., the hypothesized principle provides an independent, alternate, quantitative, structural basis for the empirical observation that $\ln(S_u/\bar{p}_c)$ varies logarithmically with OCR.

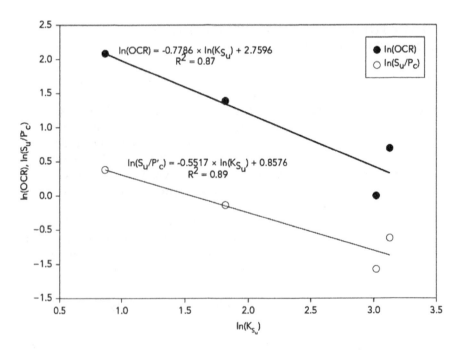

Figure 4.5 $\ln(\text{OCR})$ and $\ln(S_u/p'_c)$ vs. $\ln(K_{S_u})$ for Sheahan (1991).

For the Sheahan data, eliminating K_{s_u} from the two regression equations in Figure 4.5 gives $S_u/\overline{p}_c = 0.334 \times \mathrm{OCR}^{0.709}$ versus the measured values of $0.334 \times \mathrm{OCR}^{0.705}$ from the best fit of $S_u/\overline{p}_c$ directly against OCR. Similarly, for the Plant (1956) data, calculated $S_u/\overline{p}_c = 0.296 \times \mathrm{OCR}^{0.468}$ compares against the measured $0.297 \times \mathrm{OCR}^{0.471}$, and for the Gens (1982) data, calculated $S_u/\overline{p}_c = 0.307 \times \mathrm{OCR}^{0.768}$ compares against the measured $0.301 \times \mathrm{OCR}^{0.786}$.

To be completely honest though, it still is not clear why K_{s_u} varies as it does with OCR and $S_u/\overline{p}_c$. This is still a gap in the explanatory power of the theory. Hopefully someday, someone will find what underlying physical mechanism results in these two logarithmic relationships.

CONCLUSION

The close fits to the test data over a wide range of soils, test conditions, and strains together with the smooth variation of the curves of the model's parameters, together with its successful predictions does not falsify the hypotheses that drive the model. The model is a physical model, not a phenomenological model because it finds the root cause of observed behavior (phenomenology) in the underlying physical basis of that behavior. It is not "curve fitting" because unlike curve fitting, the parameters of the model have both direct physical meaning and scientific basis. The model's key significance is that it points to fundamental considerations underlying and driving soil shear. Its key advantage is the close fit to measured data, which in turn will result in better modeling of field problems.

The model directly complements work done in areas that more immediately investigate micromechanical soil behavior such as for ex. discrete element methods (see for example, Cundall and Stark, 1979). Hence, the finding using distinct element methods that only a minority of soil particles form a load bearing skeleton that carries the bulk of the load is in line with the suggestion in Joseph (2013a) that the particles that move to the steady-state in the Poisson process are from this minority of load bearing particles.

The advantage of the soil deformation dynamical systems model is that it is essentially describes work-hardening plasticity without the need to make any additional assumptions (which by definition must be specific and hence potentially erroneous) required to define an explicit flow rule or yield surface. As with distinct element approaches, these latter are implicit in the dynamical systems model. For example, for the Boston Blue Clay tested by Sheahan (1991), they are implicit in Figure 3.2 of Chapter 3.[1]

---

[1]One interesting finding that has emerged over the last two decades from the field of "complexity" is "the rule of hand." Gunderson and Holling (2002) write that "Being

The next chapter, studies strain-rate effects in soil shear in the light of DSSM. It concludes Part 1 of this book. However, before that, the following visualization exercises are useful to fully understand it visualize shrinking to the size of a small clay particle and are inside a soil element, next to a load-carrying particle and surrounded by other load carrying particles. Imagine what it feels like as shear proceeds and particles around move. Next, visualize a brick sliding on a table and how, regardless of vertical load on the brick, the ratio of the shear to vertical stresses stays the same. Imagine the brick at the quasi-static state (almost moving) and how the static coefficient of friction must govern. Now imagine it moving and the coefficient of friction reducing to its dynamic value. Imagine how as the rate at which it slides changes these coefficients of friction change, depending on the rate. Finally, visualize

---

as simple as possible but no simpler than necessary leads to the 'rule of hand'." The rule is counter intuitive and as best known, the first person to have sensed it was Isaac Newton, who felt strongly that nature was "… simple, and always consonant to itself." The rule implies that complicated models are not necessary to explain the key patterns of interest and, in fact, are likely to mask them. Though the bulk of the work that led to the formulation of this "rule of hand" appears to have come from lessons learned understanding and modeling social and ecological systems, the rule is now finding use in understanding behaviors seen in many other complex systems such as the brain. The rule also seems to have direct relevance to the dynamical systems soil deformation model. As described below, the DSSM model has the characteristics of a "rule of hand" which are as follows (Gunderson and Holling, 2002): a) typically, three to five key interacting components (in our model, movement at random strains to the final position and captured by the variable γ, interparticle surface contact area captured by the variable J, and the coefficient of friction as captured by the variable ϕ). This is where the rule gets its name–the number of variables is not more than the number of fingers on one hand; b) three qualitatively different speed variables (slow speed variable λ which for a given stress-path varies only slightly with soil type, medium speed variable ϕ which varies with strain-rate, and high speed variable J the interparticle contact area which varies directly and continuously with shear strain). Note that here the word "speed" is a technical term, a measure of how fast a variable changes; c) nonlinear causation (in our model, the exponential decay of the interparticle contact area) and multistable behavior (the different steady-state depending on strain-rate and void ratio); d) vulnerability and resilience that change with the slow variables – λ accounts for the different behaviors (the extent and nature of dilative or contractive behavior) of a given soil. It seems to be the "slowest" of the three variables, changing only slightly across the basic soil types for a given stress-path; e) creation of structures by (the forces on) the system (in our case, the forces are the normal and shear stresses p and q respectively) and, reinforcements of these forces from the structure created, i.e., system forces and structure interact as a dynamical system; f) spatially contagious process to generate self-organized patterns (starting at the failure plane and spreading out in either direction). These similarities of the soils model and the "rule of hand" though remarkable are but expected–an assemblage of soil particles forms a "complex" (in the strict mathematical sense of the word) system and so it makes sense that it would have the standard signature of all complex systems.

the Poisson process of radioactive decay, but backwards, i.e., radioactive particles going back into the radioactive material. Write out the equation that describes this. Imagine soil deformation the same way. Try to visualize the basis of Equations 5. Write out Equation 5a, again, not from memory but from genuine understanding.

Chapter 5

Soil shear–strain rate effects

This chapter examines the analytical solution to the DSSM model and the DSSM model itself to study strain-rate effects in soils.

ANALYTICAL SOLUTION

From the dynamical system equations for the physical model, it is possible to write $\bar{p}$ and $d\bar{p}/d\gamma$ as functions of q and obtain a single equation in terms of q as:

$$\frac{d^2q}{d\gamma^2} + C\frac{dq}{d\gamma} + K_q = 0 \tag{1}$$

where:

$$C = [\lambda_p + J_q \exp(-\lambda_q\gamma) - J_p \tan \alpha_p \exp(-\lambda_p\gamma)]$$

and

$$K = J_q \exp(-\lambda_q\gamma)[(\lambda_p - \lambda_q) + J_p(\tan \alpha_q - \tan \alpha_p) \exp(-\lambda_p\gamma)]$$

A like approach results in a like equation for $\bar{p}$ or e—i.e., the behavior of $q, \bar{p}$ or e is, in each case, analogous to that of a mass-damper-spring system with both damper and spring nonlinear. As shear strain increases, the damper approaches its corresponding constant λ value-λ_p in the case of Equation (1)-while the spring stiffness approaches zero.

Equation (1) for q and the corresponding equations for $\bar{p}$ or e are homogeneous, second-order, linear equations with non-autonomous coefficients. Such equations typically have no closed form solution and this is the case here. However, Equation (1) does have an analytical solution for the special case of equal λ_q and λ_p. Assuming they are equal for the purpose of an analytical solution does not seem unreasonable given that the rates λ_q, λ_p at which particles move into the flow-structure seem close to each other across the entire range of OCR for the triaxial test data examined to date. The detailed, manual derivation of the solution to Equation (1) is in Appendix 2-analogous derivations would apply if had Equation (1) been expressed in terms of $\bar{p}$ or e.

As detailed in Appendix 2, the analytical solution for Equation (1) is:

$$q(\gamma) = C_1 \exp(-\omega_1 \exp(-\lambda\gamma)) + C_2 \exp(-\omega_2 \exp(-\lambda\gamma)) \tag{2a}$$

$$\bar{p}(\gamma) = \frac{\lambda}{J_q \tan\alpha_q} \left[C_1\left(\omega_1 + \frac{J_q}{\lambda}\right) \exp(-\omega_1 \exp(-\lambda\gamma)) \right.$$
$$\left. + C_2\left(\omega_2 + \frac{J_q}{\lambda}\right) \exp(-\omega_2 \exp(-\lambda\gamma)) \right] \tag{2b}$$

where:

$$\omega_{1,2} = \frac{-(J_p \tan\alpha_p - J_p) \pm \sqrt{(J_p \tan\alpha_p - J_q)^2 - 4J_q J_p(\tan\alpha_q - \tan\alpha_p)}}{2\lambda} \tag{2c}$$

and C_1, C_2 are determined by the boundary conditions at the steady-state.

At the steady-state, γ is very large and from Equations (2), steady-state stresses $q(ss)$ and $\bar{p}(ss)$ are:

$$q(ss) = C_1 + C_2 \tag{3a}$$

$$\bar{p}(ss) = \frac{\lambda}{J_p \tan\alpha_q} \left[C_1\left(\omega_1 + \frac{J_q}{\lambda}\right) + C_2\left(\omega_2 + \frac{J_q}{\lambda}\right) \right] \tag{3b}$$

For a given shear test, it is possible to calculate the value of the steady-state shear and confining stresses to very large strains (typically over 50%) using the numerical Range-Katta solution to Equations (1), to where they stopped changing in (say) the fifth decimal. Using the parameters obtained for λ, α_q and J_q from the exact numerical solution calculate the corresponding values of $\omega_{1,2}$ from Equation (2c) and then using the now known steady-state stresses, solve Equations (3) for C_1 and C_2. Use these values in Equations (2a, b) to obtain the full stress-strain curves. However, because initial conditions are not well defined, the analytical solution to the approximated equation does not presently provide good fits except at large shear strains, i.e., its use at present seems limited.

STRAIN-RATE EFFECTS

Changes caused by strain-rate are small–tests on Boston Blue clay (CL) by Sheahan (1991) show that peak and steady-state stresses varied by about 10% over a range of three orders of magnitude of change in strain-rate, i.e., not large enough to influence significantly the dynamical system model's parameters for the Sheahan data. Likewise, other test data empirically evidence strain-rate effects. For example, Yamamuro *et al.* (2011) observed a 30% increase in peak shear stresses over an approximately six order of magnitude increase in strain-rate for tests on a clean sand, though for their tests at very high strain-rates of over 1500%, inertial effects, not considered in this present paper, cannot be ruled out. For summaries of strain-rate effects in soil-shear, see Yamamuro *et al.* (2011), Díaz-Rodríguez *et al.* (2009), Hossain and Randolph (2009), and An *et al.* (2011). In general, except for very high values, the strain-rate plays a small, subtle role in shear behavior as the rest of this paper explains using the framework of the dynamical systems model.

STRAIN-RATE IMPLICATIONS OF THE MODEL

Chapter 3 applied the model to tests at various strain-rates conducted by Sheahan (1991) and found that parameters correlated with OCR with high correlation coefficients, and that this was regardless of strain-rate, i.e., that behavior was effectively strain-rate independent. The other test sets (Plant, 1956 and Gens, 1982) reported in Joseph (2012) were all conducted at the same strain-rate and so could not be used to determine the effect of strain-rate. The rates λ_q, λ_p at which particles move into the flow-structure are relatively close to each other across the entire range of OCR examined. If they are assumed to be equal, the model has an analytical solution given by Equations (2). Likewise, $e(\gamma)$ has a similar solution.

At the start of plastic strain, $\gamma = 0$ and from Equation (2), stresses $q(0)$ and $\bar{p}(0)$ corresponding to the initiation of plastic strain are:

$$q(0) = C_1 \exp(-\omega_1) + C_2 \exp(-\omega_2) \tag{4a}$$

$$\bar{p}(0) = \frac{\lambda}{J_q \tan \alpha_q} \left[C_1 \left(\omega_1 + \frac{J_q}{\lambda} \right) \exp(\omega_1) + C_2 \left(\omega_2 + \frac{J_q}{\lambda} \right) \exp(\omega_2) \right] \tag{4b}$$

Note that strictly speaking, $q(0)$ and $\bar{p}(0)$ are not the initial shear and confining stresses measured at the start of a test. Rather they correspond to the values of shear and confining stresses at which plastic deformation begins. Atkinson (1993) and Hicher (1996) among others confirm intuitive expectations in that elastic behavior occurs only up to very small strains, in the order of 0.00001% or less, i.e., for all practical purposes, plastic strain (when the first grain moves to its final position), initiates at the very start of the test. Note also, that for the case of zero plastic strain ($\gamma = 0$), the rigorous version of the model as described in Equations (2) of Chapter 3 becomes:

$$\frac{dq}{d\gamma_{(\gamma=0)}} = J_q[\bar{p}_0 \tan \alpha_q - q_0] \tag{5a}$$

$$\frac{d\bar{p}}{d\gamma_{(\gamma=0)}} = J_p[\bar{p}_0 \tan \alpha_p - q_0] \tag{5b}$$

which shows that the slope of the stresses versus plastic-strain at the initiation of plastic strain are functions of $J_q, J_p, \alpha_q, \alpha_p, q(0)$, and $\bar{p}(0)$ but independent of λ_q and λ_p.

Finally, at the steady-state, γ is very large and from Equation (3), steady-state stresses $q(ss)$ and $\bar{p}(ss)$ are:

$$q(ss) = C_1 + C_2 \tag{6a}$$

$$\bar{p}(ss) = \frac{\lambda}{J_p \tan \alpha_q}\left[C_1\left(\omega_1 + \frac{J_q}{\lambda}\right) + C_2\left(\omega_2 + \frac{J_q}{\lambda}\right)\right] \tag{6b}$$

At very high strain-rates where deformation velocities are sufficiently high, inertial effects of particles including inter-particle collisions occur. As always this book does not take inertial effects into account because the only behavior to which the DSSM theory as presently defined applies is that which occurs when strain-rates are not high enough to cause inertial effects.

APPLICATION

To compare tests at different initial consolidation stresses at a given OCR, normalize shear and normal stresses by σ'_{vc} the effective vertical consolidation stress. For each test, calculate the value of these normalized stresses to very large strains (typically over 50%) using the rigorous, numerical solution to Equations (2), to where they stop changing in (say) the fifth decimal. Use these values, the fitted parameters for the test, and the values of $\omega_{1,2}$ from Equation (3c) to solve Equations (6) to obtain C_1 and C_2. Next use these values in Equations (4) to calculate $q(0)$ and $\bar{p}(0)$.

Figure 5.1 plots these calculated stresses at the start of plastic strain against the test strain-rate for tests by Sheahan (1991) on samples with an OCR of 1.

The data show that the stresses at the start of plastic strain, $q(0)$ and $\bar{p}(0)$, vary with the strain-rate. Because of rapidly changing values at the start of a shear test, initial conditions are hard to measure accurately and Sheahan (1991) reported slightly negative initial stresses for tests at an OCR of 1 at very low strain-rates. Figure 5.2 plots the extrapolated steady-state values $q(ss)$ and $\bar{p}(ss)$ as obtained from Equations (6).

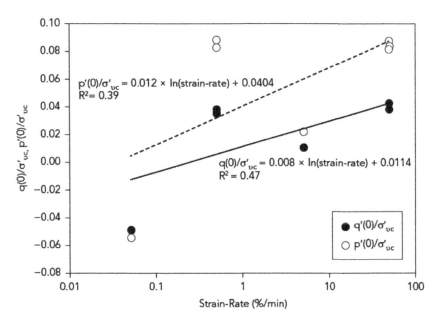

Figure 5.1 Variation of normalized $q(0)$ and $\bar{p}(0)$ with strain-rate for tests with OCR = 1 from Sheahan (1991).

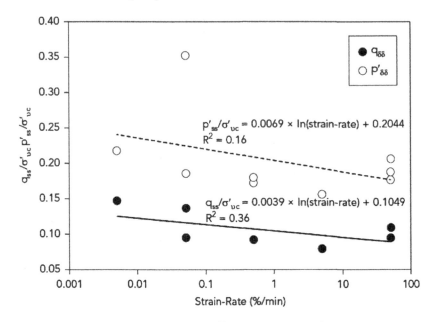

Figure 5.2 Variation of normalized $q(ss)$ and $\bar{p}(ss)$ with strain-rate for tests with OCR = 1 from Sheahan (1991).

Table 5.1 Variation of initial stresses, initial slopes, maximum, and steady-state stresses with strain-rate from Sheahan (1991) compression tests.

OCR	$\dfrac{dq_0}{d(\ln\dot\gamma)}$	$\dfrac{d(dq/d\gamma_0)}{d(\ln\dot\gamma)}$	$\dfrac{dq_{max}}{d(\ln\dot\gamma)}$	$\dfrac{dq_{ss}}{d(\ln\dot\gamma)}$	$\dfrac{d\bar p_0}{d(\ln\dot\gamma)}$	$\dfrac{d(d\bar p/d\gamma_0)}{d(\ln\dot\gamma)}$	$\dfrac{d\bar p_{max}}{d(\ln\dot\gamma)}$	$\dfrac{d\bar p_{ss}}{d(\ln\dot\gamma)}$
1	0.01	0.10	−0.01	−0.003	0.01	0.11	−0.03	−0.01
2	0.01	0.03	0.01	0.02	0.02	0.05	0.02	0.05
4	0.12	0.01	0.01	0.02	0.32	0.15	0.02	0.02
8	−0.11	−0.08	0.02	0.02	−0.30	−0.02	0.04	0.05

The data from the confining stress values for tests at a strain-rate of 0.06% and plotted in both figures were outliers, but nonetheless, included in the regression analysis. The low correlation coefficients reflect the dependence of the calculations behind both Figures 5.1 and 5.2 on the initial values of the test, values which are both low in magnitude and hard to measure accurately because of the rapidity with which conditions change at the very start of a test. Table 5.1 presents results of this same analysis for the remaining Sheahan tests conducted for higher OCR values. In general the change with strain-rate of the normalized peak and steady state stresses is very small at all OCRs (generally less than 2% and 5% per order of magnitude change of strain-rate for the normalized shear and confining stresses respectively), and smallest for the normally consolidated case.

LaGatta (1970, 1971) studied the effect of the rate of displacement on the residual strength of clays and shales, using a simple rotation shear machine that was capable of a wide range of shear velocities. He ran tests on annular specimens of remolded clay over a wide range of strain-rates. Specimens had an outside diameter of 7.11 cm, an inside diameter of 5.08 cm, and an initial thickness before consolidation of 2 mm. LaGatta's reported only initial conditions at the start of shear and final conditions at the steady-state and consequently, Equations (2) are not solvable. However it is possible to directly examine how LaGatta's measured steady-state stresses vary with strain-rate. Additionally, LaGatta (1971) reports the results of steady-state shear tests conducted on London clay and Bearpaw shale over a wide range of strain-rates. The London clay, known colloquially as "Blue London clay," had a natural water content of 28% and liquid and plastic limits of 72 and 22 respectively, which classified it as CH per the USCS. The Bearpaw shale was a grayish black, firm, clay, had a natural water content of about 33% and liquid and plastic limits of 83 and 24 respectively, which classified it also as CH per the USCS. Figure 5.3 plots the steady-state shear stress values for the London clay and the Bearpaw shale against deformation rate. Again, notice that the trend in the steady-state stresses with a very large range of shear velocities is very small.

Additional data comes also from Lagatta (1970) from tests he conducted at different shear-velocities on Pepper shale, crushed, air-dried Cucaracha

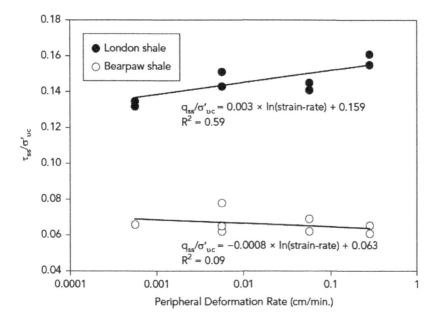

Figure 5.3 Variation of normalized steady-state strengths with deformation rate from LaGatta (1971).

shale, and the London clay. The Pepper shale was a hard, brittle clay at a natural water content of about 21%, a liquid limit of 72 and a plastic limit of 22, classifying as CH per the USCS. The Cucaracha shale was a hard, friable material, with a liquid limit of 49 and a plastic limit of 28, classifying it also as CH per the USCS. He first defined the residual shear stress at a certain shear velocity, which he then reduced by a factor of 10. In all cases, he found no significant change in the residual shear strength. In the case of the Cucaracha shale, he found that an increase in the shear velocity by a factor of 100 resulted in only a 3.5% change in the measured residual shear strength. Table 5.2 presents his key test data.

RESULTS

This section presents five results based on the data from these tests, analyzed from within the framework of the dynamical systems model.

i) Conditions at the start of plastic strain are strain-rate dependent

One cause of the small variations with strain-rate seen for the stress-strain curves would be dependence on strain-rate, of the initial values of the stresses

Table 5.2   Effect of rate of displacement on steady-state strength
(LaGatta, 1970).

Material	σ_v (kN/m²)	$\dot{\delta}$ (10⁻³ m/min)	τ_{ss} (kN/m²)
Pepper	98.1	5.6	13.8
Shale		0.56	14.9
Crushed	784.5	56	99.5
Cucaracha		5.6	96.1
Shale		0.56	96.1
London	392.3	5.6	56.1
Clay		0.56	55.0

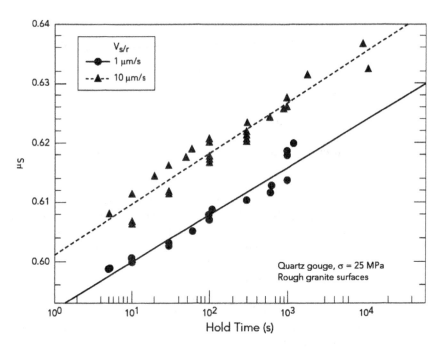

Figure 5.4 Variation of the static coefficient of friction with applied velocity and hold time
(from Marone, 1998b).

at the start of plastic-strain. Such a dependence on strain-rate of the stresses
prior to the start of plastic shear deformation would help explain any subsequent differences with strain-rate in the remaining (the majority) portion of
the stress-strain curve even if the operative mechanisms in that region were
themselves insensitive to strain-rate. Bowden and Tabor (1950) and Marone
(1998a) have, among others, shown that the static coefficient of friction
varies with the initial applied velocity (see Figure 5.4 for an example), and

hence provide a fundamental basis for why initial conditions would vary with strain-rates.

Figure 5.1 which is a plot of the stresses $q(0)$ and $\bar{p}(0)$ at the start of plastic strain against strain-rate for Sheahan's (1991) data shows that this dependence does appear to exist ensuring that the initial slope of the stress-strain curves will also depend on the strain-rate. Table 5.1 for the Sheahan (1991) data shows that these initial slopes at the start of plastic strain do vary with strain-rate. Yamamuro *et al.* (2011) also report such a dependence.

Hence, the first observation, based on frictional considerations deriving from dynamical systems soil shear theory, with preliminary confirmation by test data: conditions (initial stresses and the initial slope of the curves) at the start of plastic deformation are strain-rate dependent.

ii) Linear behavior does not mean elastic behavior

Strain-rate dependence of the stresses prior to plastic deformation would in turn be due to any dependence of the static coefficient of friction on the rate of application of stresses or due to the soil behaving as a viscoelastic solid. The solid lines in Figure 5.5 shows measured data from a typical shear test by Castro (1969) on a sand compacted to a point well below the steady-state line so that it would dilate during shear. Of interest is the early region (less than 0.5% axial strain) of the deformation curves. q and $\bar{p}$ (both measured and calculated) vary linearly in the early part of the curve. The traditional explanation for this initial linear deformation is that it is elastic (see for example, Hirshfeld, 1958, and Wroth and Bassett, 1965).

The measured void-ratio curve shown labeled in Figure 5.5 (plotted against the Y-axis on the right for clarity) is also typical for dilative materials. Initially, the void-ratio decreases, before the sample starts to dilate at about 0.5% strain. Arulmoli *et al.* (1992) and Shapiro (2000) ran numerous drained tests on "Nevada sand" and "Bonnie silt" (Figure 5.6) and found this very visible contraction in the early part of the curve for their dilative samples. Again, the traditional interpretation for this decrease in void-ratio is that it is due to initial elastic deformation during which the sample contracts, after which, once particles fully engage, plastic deformation begins with the expected dilation as the sample moves to the steady-state.

The model applied to both Castro's data and Arulmoli *et al.*'s data provide close fits for not just q and $\bar{p}$, but also for the early dip observed in the void-ratio curves. Since the model does not describe elastic strains, the close fit to this dip in the void-ratio curve implies that even at near zero strains the soil is behaving in the same manner as the rest of the curve, i.e., is deforming plastically due to particles moving into the steady-state flow structure at random shear strains.

The behavior of the model (Equations 2 in Chapter 3) for small values of the shear strain γ explains the early variation of q and $\bar{p}$. For the test

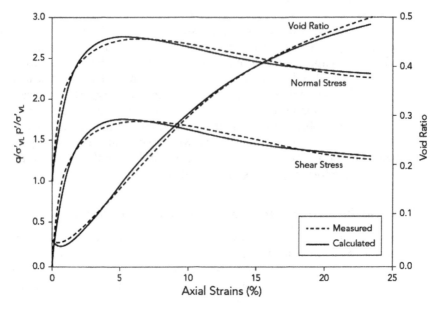

Figure 5.5 Test 1-5 data from Castro (1969) fit by Joseph (2012) model.

analyzed in Figure 5.5, the λ_q and λ_p values are about 0.08. This means that at strains of 0.5%, the value of $\exp(-\lambda_q\gamma)$ or $\exp(-\lambda_q\gamma)$ is about 0.985 or close to 1 and the equations that define the model for these strains or less become:

$$\frac{dq}{d\gamma} = J_q[\bar{p}\tan\alpha_q - q] \tag{7a}$$

$$\frac{d\bar{p}}{d\gamma} = J_p[\bar{p}\tan\alpha_p - q] \tag{7b}$$

Equations (7) represent an ordinary, second order, linear differential equation with constant coefficients whose solution is:

$$q(\gamma) = C_1\exp(-\omega_1\gamma) + C_2\exp(-\omega_2\gamma) \tag{8a}$$

$$\bar{p}(\gamma) = \frac{\lambda}{J_q\tan\alpha_q}\left[C_1\left(\omega_1 + \frac{J_q}{\lambda}\right)\exp(\omega_1\lambda) + C_2\left(\omega_2 + \frac{J_q}{\lambda}\right)\exp(\omega_2\gamma)\right]$$

$$\tag{8b}$$

From Equations (8), for small strains:

$$\frac{dq}{d\gamma} = -C_1\omega_1 - C_2\omega_2 \tag{9a}$$

$$\frac{d\overline{p}}{d\gamma} = \frac{\lambda}{J_q \tan\alpha_q}\left[C_1\omega_1\left(\omega_1 + \frac{J_q}{\lambda}\right) + C_2\omega_2\left(\omega_2 + \frac{J_q}{\lambda}\right)\right] \tag{9b}$$

In other words, for small strains, the initial slope of the stress-strain curves is constant, i.e., the curves themselves are initially linear. This matches with observations from test data that show that the initial portion of the stress-strain curves for q and $\overline{p}$ are linear. As seen above, this linearity is a result not of elastic behavior, but rather, the linearization that occurs at small strains due to the small effect of the nonlinear terms in the equations that describe deformation for the rest of the curve.

Mathematically, it happens to be, that for the case of soil shear, the intrinsically non-linear stress-strain curve is nearly linear at small strains. The reason for this is that at small strains the underlying physical behavior as described by the governing equations is approximately linear. The analysis above shows that for the case of the equations describing soil shear that linearity exists at trains as high as 0.5%. As mentioned earlier, Atkinson (1993) and Hicher (1996) among others report that elastic behavior occurs only up to very small strains, in the order of 0.00001% or less. Beyond this, even though the stress-strain curves appear linear, the damage to the soil structure that has already occurred ensures that the soil will not return to its original state on unloading, i.e., even in this early region of the curve, plastic deformation has begun.

These considerations lead to the second result: the linear nature of the early part of a stress-strain curve does not mean behavior is elastic. Rather, it is the expected linear behavior at small strain of the same mechanism of non-linear plastic deformation that governs the entire remaining stress-strain curve–the Poisson process of particles moving into the steady-state flow structure at random shear strains.

iii) Role of static coefficient of friction in initial decrease in void-ratio

In 1758, Johann Andreas von Segner was the first known scientist to draw a distinction between static and dynamic friction, a distinction that Coulomb accounted for in his friction law (Dowson, 1997). More recently, Bowden and Tabor (1954) and Marone (1998a) among others, have shown that for most materials, the dynamic coefficient of friction is less than or equal to the static coefficient of friction, and that for many materials, all it takes is movement of a fraction of a millimeters or less for the static friction value to decrease to that of dynamic friction.

An examination of the equation for void ratio change (Equation 2c in Chapter 3) shows that a high initial static coefficient of friction can result in the initial effects of the confining stress exceeding those of the shear stress, thereby causing the initial dip in the void-ratio curve. This high initial value of the coefficient of friction decreases with very little deformation, to its lower dynamic value. Once the dynamic coefficient of friction dominates, then the effect of the shear stresses begins to exceed those of the confining stresses, and the soil dilates. The change from static to dynamic friction also causes a slight break in the slope of the stress-strain curves. However, because compared with their absolute values the net effect on the stresses is small, the break is often not clearly visible; for example, even in Figure 5.5, it is hardly visible.

This leads to the third result: static friction governs soil behavior prior to the start of plastic deformation. This static coefficient of friction is larger than the dynamic friction coefficient. The change from the larger static friction coefficient to a smaller dynamic friction coefficient causes the small, early breaks in the stress-strain and void ratio curves seen in materials that dilate when sheared. A model that accounted for the static coefficient of friction acting at small strains would show yet better fits than those shown in Figures 5.5 and 5.6.

iv) Steady-state stresses vary only slightly with strain-rate

By definition, the steady–state condition requires that all particles be in the statistically constant steady-state flow structure, i.e., at the steady-state, shear occurs on a statistically constant inter-particle surface contact area. Hence, any sensitivity of the steady-state stresses to strain-rate is due to the sensitivity of the dynamic coefficient of friction of the particle material to strain-rate. Observe from Figures 5.2 and 5.3 that steady-state stresses for the Boston Blue clay, the Bearpaw shale, and the London clay, vary a very small amount over a large range of strain-rates. Additional data from tests on Pepper shale, Cuaracha shale, and the London clay (LaGatta, 1970) taken to the steady-state at different strain-rates (see Table 5.2) likewise confirm this very small variation with strain-rate of the steady-state stresses.

This leads to the fourth result: variation in steady-state stresses are very small, and primarily attributable to the slight variation of the dynamic coefficient of friction with strain-rate. For all practical purposes, steady-state stresses are independent of strain-rate.

v) Dynamic friction's strain-rate dependence effects plastic deformation

With the start of plastic deformation, the coefficient of dynamic friction applies, and from here on, the soil is sensitive to strain-rate only as far as the

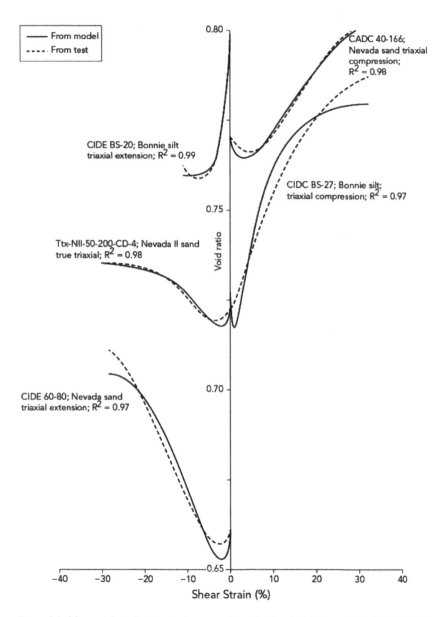

Figure 5.6 Measured void ratio-strain curves from Arulmoli (1992) and Shapiro (2000) fit by Joseph (2012) model.

dynamic coefficient of friction is. As seen previously from the results from Sheahan (1992) and LaGatta (1970, 1971), the dependence on strain-rate of the steady-state stresses is small. Table 5.1 also shows the variation of the peak shear stresses with strain-rate for the tests by Sheahan. Notice that this variation is very similar to that of the steady-state values. Since the structure obtaining at the steady-state is the flow structure, it appears that the different structure obtaining at the peak stress values has little to no effect on the variation of peak stresses with strain-rates.

This leads to the fifth and final result namely that once plastic deformation begins, any variations in stress-strain and void-ratio-strain curves with strain-rate are primarily due to the dependence of the dynamic coefficient of friction on strain-rate. It is worth noting though that at extremely high strain-rates (above 1500% strain/second), Yamamuro *et al.* (2011) saw anomalous behavior: at peak shear stress levels the peak stresses did not correlate with the maximum dilatancy rate. They attributed this to shear banding preventing failure at traditional maximum dilatancy rates. However it is likely that for their tests at these very high strain-rates, inertial effects were coming into play.

CONCLUSION

Analysis in the context of DSSM suggests that plastic deformation dominates from almost the very start of shear and that strain-rate effects are due primarily to dependence of the static and dynamic coefficients of friction on strain-rate. This is expected given the fundamental role of simple friction in shear. Friction coefficients depend very slightly on strain-rate; so, soil shear also depends very slightly on strain-rate.

The findings are tentative given the small sample size studied in the paper and need more tests like Sheahan (1991) to confirm them. In particular, the initial stages of a test are very important for strain-rate related studies and special measures (voluminous high-speed data recording) are required to determine all-important initial conditions accurately. These initial conditions are key, determining the exact location of the remainder of the stress-strain curves. Nonetheless, the present data set appears to confirm the hypothesis that the coefficients of friction explain strain-rate related observations.

It is of interest to note that this model of static friction moving to dynamic friction during continued shear is in line with an influential paradigm for stick-slip seen on rock surfaces and rock-gouge-the Dieterich-Ruina model (see for example Scholz, 1998). In this paradigm, the static coefficient of friction controls initial slip. Upon very little displacement, this static coefficient of friction rapidly decreases to a lower value-the dynamic coefficient of friction-which then controls subsequent movement.

Chapter 6

Consolidation compression–the generalized model

This chapter generalizes the dynamical systems soil shear model to describe deformation along any stress path. It is based on Joseph (2014a).

The generalization states that for any stress path, particles move at random shear strains into the final structure corresponding to the applied stresses. In short, a friction based dynamical system describes all soil deformation including, one-dimensional consolidation, and, not just shear to failure and beyond to the steady-state condition.

The model remains the same and is as follows:

$$\frac{dq}{d\gamma} = J_q[\overline{p}\tan\alpha_q \exp(-\lambda_p\gamma) - q\exp(-\lambda_q\gamma)] \tag{1a}$$

$$\frac{d\overline{p}}{d\gamma} = J_p[\overline{p}\tan\alpha_p \exp(-\lambda_p\gamma) - q\exp(-\lambda_q\gamma)] \tag{1b}$$

$$\frac{-de}{d\gamma} = J_e[\overline{p}\tan\alpha_e \exp(-\lambda_p\gamma) - q\exp(-\lambda_q\gamma)] \tag{1c}$$

where: e = void-ratio, q = shear stress, $\overline{p}$ = effective normal/confining stress, J_q, J_p = initial non-dimensional inter-particle contact area of the load bearing particles that are not in the final structure, per unit area of the sample, corresponding to q and $\overline{p}$ respectively, J_e = void-ratio due to load bearing particles not being in the final structure, per unit stress, per unit area, α_q, α_p, α_e = secant friction angles corresponding to q, $\overline{p}$, e respectively, γ = shear strain, and λ_q, λ_p = rate of movement of particles into the final structure, corresponding to q and $\overline{p}$ respectively, per unit of strain.

To confirm this generalization, this chapter will specialize the model to the particular case of one-dimensional consolidation, which, it is very important to realize, is nothing but shear along a K line; for the particular case of one-dimensional normal consolidation (NC), it is shear along the K_0 line. Specializing the generalized model to this case results in its predicting the well-established linear relationship between void-ratio e and the logarithm of the vertical effective stress $\overline{\sigma}_v$ for normal consolidation, thereby confirming the generalized model.

A close reading of the literature shows some empirical evidence for a power law relationship between e and $\overline{\sigma}_v$. However, it is not clear that the difference from the linear relationship between e and $\log \overline{\sigma}_v$ seen in the test data from consolidation tests taken to high vertical effective stresses is not due to grain crushing (see Butterfield, 1979, Cargill, 1984 and Townsend, 1987 among others). A comparison of the two types of curves shows that at high stresses, the power law curve has lower values of void-ratio that the corresponding semi-log equivalent. This is as expected if grain crushing is occurring–the now smaller grains pack together better and so exhibit lower void-ratios (and a power-law relationship) at higher stresses.

As described in Joseph (2014a), as was the case for shear to failure, so also for any stress path—soil microstructure changes in a Poisson process, i.e., particles move at random shear strains into the final structure corresponding to the applied stresses. In other words, movement of the soil particles into the final structure for any stress-path occurs as a random event independent of the strain for which the particle was not in this structure. This means that the model in Chapter 3 for the friction based dynamical system for shear taken to the steady-state, apply directly, even for one-dimensional consolidation.

While the model in Chapter 3 was explicitly for the case of the final structure being the steady-state flow structure, the generalization of the model extends this behavior to any final structure, including that resulting from one-dimensional consolidation. The parameters of the dynamical system equations take on values appropriate to the stresses acting during soil deformation. As with shear to the steady-state condition, so too for the case of one-dimensional consolidation: load-bearing particles that have not yet moved into the applicable final structure determine behavior and parameter values.[1] The next section specializes the model for this case.

SPECIALIZATION FOR ONE-DIMENSIONAL CONSOLIDATION

For one-dimensional consolidation, the following well-known, standard relationships hold for the principle stresses:

$$\overline{\sigma}_1 = \overline{\sigma}_v$$

where $\overline{\sigma}_v$ is the vertical effective stress and $\overline{\sigma}_2 = \overline{\sigma}_3 = \overline{\sigma}_h = K\overline{\sigma}_v$ where $\overline{\sigma}_h$ is the horizontal effective stress and K is the ratio of the horizontal to vertical

---

[1] For simplicity, the implicit assumption made in the derivation in this chapter is that no grain crushing is taking place under loading. The DSSM concept probably holds even if grain crushing does occur, i.e., the crushed particles will behave like the other particles as well, moving to their final position at random shear strains.

stress for one-dimensional conditions. For one-dimensional normal consolidation, $K = K_0$ the coefficient of lateral earth pressure at rest. Replacing $\bar{\sigma}_1$ with $\bar{\sigma}_v$ and $\bar{\sigma}_2, \bar{\sigma}_3$ with $K\bar{\sigma}_v$, octahedral stresses q and $\bar{p}$ become:

$$q = \frac{\sqrt{2}(1-K)\bar{\sigma}_v}{3} \quad \text{and} \quad \bar{p} = \frac{1+2K}{3}\bar{\sigma}_v$$

Octahedral shear strain γ for one-dimensional compression, is $\gamma = \frac{\sqrt{2}}{3}\varepsilon_v$ and consequently, $d\gamma = \frac{\sqrt{2}}{3}d\varepsilon_v$, and $\frac{dq}{d\gamma}, \frac{d\bar{p}}{d\gamma}$ become:

$$\frac{dq}{d\gamma} = (1-K)\frac{d\bar{\sigma}_v}{d\varepsilon_v} \quad \text{and} \quad \frac{d\bar{p}}{d\gamma} = \frac{(1+2K)}{\sqrt{2}}\frac{d\bar{\sigma}_v}{d\varepsilon_v}.$$

Substitute these into Equations (1) to get two expressions for $\frac{d\bar{\sigma}_v}{d\varepsilon_v}$.
   From Equation (1a):

$$\frac{d\bar{\sigma}_v}{d\varepsilon_v} = \frac{J_q}{3(1-K)}[(1+2K)\tan\alpha_q \exp(-\lambda_p\varepsilon_v)$$
$$- \sqrt{2}(1-K)\exp(-\lambda_q\varepsilon_v)]\bar{\sigma}_v \tag{2a}$$

and from Equation (1b):

$$\frac{d\bar{\sigma}_v}{d\varepsilon_v} = \frac{\sqrt{2}J_p}{3(1+2K)}[(1+2K)\tan\alpha_p \exp(-\lambda_p\varepsilon_v)$$
$$- \sqrt{2}(1-K)\exp(-\lambda_q\varepsilon_v)]\bar{\sigma}_v \tag{2b}$$

Equate 2a and 2b, and simplify to express $\exp(-\lambda_q\varepsilon_v)$ in terms of $\exp(-\lambda_p\varepsilon_v)$ as:

$$\exp(-\lambda_q\varepsilon_v) = const1_{ocr} \exp(-\lambda_p\varepsilon_v) \tag{3}$$

where:

$$const1_{ocr} = \left[\frac{\dfrac{J_q \tan\alpha_q(1+2K)}{(1-K)} - \sqrt{2}J_p \tan\alpha_p}{\sqrt{2}\left(J_q - \dfrac{\sqrt{2}J_p(1-K)}{(1+2K)}\right)} \right]$$

is a coefficient that is a constant for a given OCR (terms $J_q, J_p, \alpha_q, \alpha_p, K$ are properties that depend only on OCR). For the case of normal consolidation $K = K_0$ and correspondingly, let $const1_{ocr}$ now simply be $const1$.

Substitute Equation (3) into Equation 1c and simplify, to get for the case of one-dimensional normal consolidation:

$$-\frac{de}{d\varepsilon_v} = const2 \, \exp(-\lambda_p\varepsilon_v)\overline{\sigma}_v \tag{4}$$

where:

$$const2 = \frac{\sqrt{2}J_e}{9}[(1 + 2K_0)\tan\alpha_e - \sqrt{2}(1 - K_0)const1]$$

For one-dimensional normal consolidation, for an initial void-ratio e_0 at the start of the test, $\varepsilon_v = \frac{e_0-e}{1+e_0}$, i.e., $d\varepsilon_v = \frac{-de}{1+e_0}$, or $-\frac{de}{d\varepsilon_v} = 1 + e_0$. Substitute in Equation 4 to get:

$$1 + e_0 = const2 \, \exp\left[-\lambda_p\left(\frac{e_0 - e}{1 + e_0}\right)\right]\overline{\sigma}_v \tag{5}$$

Solving Equation 5 for e:

$$e = const3 - C_c \ln\sigma_v \tag{6}$$

where: $const3 = e_0 - \frac{1+e_0}{\lambda_p}\ln\frac{const2}{1+e_0}$ the void-ratio corresponding to zero vertical stress and $C_c = \frac{1+e_0}{\lambda_p}$ is a constant, the compression index along the normal consolidation line.

In short, as for the case of shear to the steady-state, so also for one-dimensional consolidation—particles move at random shear strains into the final K structure corresponding to the final applied effective vertical stress. The model predicts that the result of this stochastic behavior is the standard linear relationship between e and $\log\overline{\sigma}_v$, empirically found to exist for normal consolidation. In other words, it is possible to derive directly from the general model the fundamental, well-known, empirically evidenced relationship between e and $\log\overline{\sigma}_v$, thereby confirming the generalized model with its underlying hypothesis that particles move to their final position at random shear strains.

The next chapter examines "viscosity and creep effects" in the light of DSSM. Before that, though consider these questions: does an exponential decay ever stop? Is it possible to visualize an exponential decay continuing to infinite strain (for soils) or time (for radioactive materials)?

Chapter 7

One-dimensional loading–strain rate effects

Chapter 5 examined strain rate effects when shearing a soil and chapter 6 how dynamical systems theory also applies to one-dimensional consolidation, because one-dimensional consolidation is but shear along a K line and the same root cause–the Poisson process–drives deformation. This chapter applies this knowledge to something that has been a source of controversy for the last four decades. The work described in this chapter largely comes from Joseph (2014b).

The controversy concerns strain rate (viscosity) effects seen in one-dimensional consolidation tests. For static loading in a one-dimensional consolidation oedometer test, Gholamreza Mesri and his coworkers were the first to suggest that the plot of the void ratio immediately at the end of primary consolidation (not the 24-hour value) against the log of the vertical stress (called the End Of Primary curve or EOP curve for short) was unique. Mesri also showed that the ratio of the creep-coefficient to the compression index $(C_{(\alpha e)}/C_c)$ was constant for a given soil, and further, this ratio seemed to change very slightly across all clayey soils in general. With the advent of constant-strain one-dimensional consolidation testing, the question arose: what is the effect of the strain-rate? Does it cause the EOP curve to be non-unique and consequently does the H^2 scaling law really apply when strain-rates in the field are often so different from those used in laboratory testing. Do these viscous effects also effect secondary consolidation?

This chapter explains how dynamical systems soil-mechanics theory suggests that strain rate (viscosity) effects and secondary consolidation each have a separate physical basis. As shown in Chapter 5, viscous behavior is due to strain-rate dependence of the coefficients of friction at interparticle contacts as they slide against each other in simple friction, and so, should occur during both primary consolidation and secondary consolidation. As also shown in Chapter 5, these effects are small so long as shearing velocities are small enough to not cause inertial effects, i.e., are "pseudo-static."[1]

Secondary consolidation is the continued deformation of the soil structure after excess pore-pressure has dissipated, as small numbers of particles move at random shear strains, in a Poisson process, to new final positions. As detailed in this chapter, a root cause analysis based on the Poisson based process reveals that a) $C_{(\alpha e)}/C_c$ must be constant for a given soil and b) that this ratio is approximately constant for most soils due to the form of the equation that expresses this ratio and because very few particles move to new positions during secondary consolidation.

The essential point is this: strain-rates for typical geotechnical field problems being pseudo-static, viscous effects in the field are small and the current design practice of using the EOP curve and $C_{(\alpha e)}$ to calculate compression settlements appears reasonable unless the soil is very compressible and the soil layers very thick. The empirical evidence from different soils and stress-paths suggests that adsorbed water layers control strain-rate related effects.

BACKGROUND

When a saturated soil layer consisting of fine-grained soils is subject to one-dimensional loading, the resulting deformation occurs in two successive phases, primary consolidation and secondary consolidation. During primary consolidation phase, the low permeability of the soil structure restricts pore fluid drainage. Consequently, pore pressures increase and the load transfers to the pore fluid. These excess pore pressures dissipate with time as the sample drains, and the initial total stress applied, becomes effective over time, as it transfers from the pore fluid to the soil structure. At the end of primary consolidation (EOP), excess pore pressures have dissipated and the initial total applied stress is fully effective. Thereafter the soil continues to deform,

---

[1]Note: It is very important that you really understand this term "pseudo-static." While it sounds very grand, all it means is this–particles move at a rate slow enough so that there are no "inertial effects." In other words, the shear velocities are slow enough to where each particle has only negligible momentum such that if one particle encounters another, the encounter is not a "collision" with energy transfer between the colliding particles. This is important because if there are inertial impacts, then the problem becomes extremely complicated and no longer expressible in terms of the simple dynamical systems equations of the DSSM model. Fortunately, geotechnical problems usually are pseudo-static, at least to start with. Huge rockslides, once in full flow, are no longer pseudo-static, but at this stage, it is too late anyway, the safety under the initial pseudo-static condition was not sufficient and we can't do any pseudo-static based design for a non-pseudo-static condition. In other engineering problems, such as the fluidized bed of certain furnaces, conditions are no longer pseudo-static and the problem becomes extremely complex. For an example of an attempt to describe a fluidized bed furnace, see Karamavruç and Clark (1996).

but at a rate slow enough to where the fluid can drain and so, does not build up excess pore pressures, a phase referred to as secondary consolidation. During this phase, the soil continues to deform under constant effective stresses. In short, primary consolidation and secondary consolidation are two successive phases; a point marking the transition between the two phases is the EOP state.

Per Terzaghi's theory of consolidation, the time for primary consolidation is proportional to H^2 where H is the drainage length of the soil under consolidation. Standard practice in geotechnical engineering is to use the one-dimensional consolidation test conducted in the laboratory using thin samples of the order of 20 mm thickness to predict the in-situ field settlement of thick clay layers. For this reason, predictions of field behavior need to consider the scale effects due to the differing thicknesses of the laboratory sample compared with the in-situ clay deposits. Key to these scale effects is the question of the effect of soil-structure viscosity and whether it effects the deformation rate of the soil skeleton. If it did, then this would complicate scale effects.

Ladd et al. (1977) and Jamiolkowski et al. (1985) have pointed out and discussed two extreme possibilities, referred to as Hypothesis A and Hypothesis B. Per Hypothesis A creep compression strain due to viscous effects occurs only after the end of primary consolidation (EOP) and so, that the law of H^2 applies, and that the strain at EOP is the same in both thick layers in-situ, and thin layers in the laboratory test. Hypothesis B on the other hand assumes that creep compression strain due to viscous effects occurs both during primary consolidation and secondary consolidation and that so, the law of H^2 does not apply because compressive strain at EOP for a given stress is also a function of strain-rate (Leroueil 2006, Watabe and Leroueil, 2012).

To account for possible strain-rate effects, Šuklje (1957) proposed the "isotache" concept ("iso" for "equal" and "tache" for "rate") according to which there exists a unique relationship between effective stress, strain, and strain-rate that he illustrated for one-dimensional loading. Over the past 50 years there has been considerable research done on Suklje's idea—see for example Leroueil et al., 1986; 1988; Yin et al., 1994; Adachi et al., 1996; Kim and Leroueil, 2001; Hawlader et al., 2003; Imai et al., 2004; Tanaka et al., 2006; Watabe et al., 2008; 2012; Qu et al., 2010, Degago et al., 2011; Watabe and Leroueil 2012.

Nonetheless, in 2016, about six decades later, the issue stayed unresolved as to whether the EOP line is truly unique and if the H^2 scaling law applies (see for example Mesri, 2003, 2009, Mesri and Godlewski, 1977, Mesri and Choi, 1985, Watabe et al., 2008, 2009, 2013, etc.). This chapter shows that the reason for this lack of resolution is that current approaches to this problem are "phenomenological," i.e., approach the problem only in terms of the observed behavioral effects of secondary consolidation and viscous effects, and not in terms of the underlying physical mechanisms responsible for secondary consolidation and viscous effects. Finding the underlying physical mechanisms responsible for the viscous effects will resolve issues of whether the EOP line is unique and whether the H^2 scaling law applies.[2]

---

[2]This is the power of driving phenomena down to their root, physical cause. When dealing with mere phenomena, i.e., observed behavior, multiple interpretations are possible. However, analyzing phenomena down to their physical bases, severely limits the number of interpretations possible. At the shallow level of phenomena, the discussion (as in this case) can go on for decades. Every Tom (Jane), Dick (Amy), or Harry (Harriet) will have an opinion, which, since it rests on mere phenomena, is not easy to falsify directly. Every "school of thought" will be able to provide phenomenal data that match their opinion. However, once you have a hypothesis based on a physical mechanism, a hypothesis that you are not able to falsify, then suddenly, you have raised the bar ... any alternate theory must also be physically based as well as give better predictions to replace the current non-falsified theory. Continuing to discuss it merely at the phenomenal level no longer suffices. One question I receive is this: "... as an engineer the underlying physical process, i.e., the root cause itself comes second to the observed phenomena–why does the physical basis matter so much?" My answer is this: I think that the truly good engineers take the exact reverse of the first part of this statement to be true, i.e., that process (physical basis or root cause) is key/primary and that behavior is but secondary. The reason is simple–behavior (one or more phenomena) is but the manifestation of an underlying process; a given phenomenon may be driven by multiple processes but until the precise process (root cause) is identified, the best solution or stance to the problem will not be found." If I may give an anecdote about this: in my teens and twenties, my brother and I used to modify motorcycles for road racing. None of us actually raced–I was the "theoretician," my brother the "implementer," while some local lad would road race the actual machine. The machines in those days were frail bikes, whose original designers never thought that teenagers would someday "soup" them up for road racing and so in general, they could not handle the power. One problem was the "tank slapper" where suddenly, for no apparent reason, the bike would suddenly wobble under high acceleration, and seemingly, this could happen at random. The other racing teams solved this by fitting a hydraulic device to the steering. Now, while this "dampened" the "tank slapper" it a) did not make steering the bike easy for the driver and b) did not completely resolve the problem–just (literally) damped it down. Introspection revealed to me that the root cause of this problematic behavior, i.e., the physical basis of this phenomenon–was the following: two struts, one on either side of the wheel, attached the rear wheel to the main body of the motorcycle. The moment the driver released the "clutch" and applied power from the souped up engine, then this power came through a chain and pulled on a gear sprocket attached to one side of the rear wheel. I found that the two struts that attached the rear wheel to the rest of the body of the bike to be very weak and that the strut on the side where the power was applied would flex relative to the strut on the other side, at the very moment the driver released the clutch and applied power. This flexing of one strut with respect to the other pulled one end of the wheel axle (the end with the sprocket where the chain applied the power) ahead of the other end, causing the rear wheel to momentarily point slightly away from the "front, dead center" and thereby throwing the machine into a "tank slapper." I am not sure if my explanation above is clear–but anyway, the solution was to simply weld a small piece of metal to better connect the two struts that held the wheel axle, thereby greatly reducing the ability of one strut to flex with

As always, it is important to note that the strain-rates of concern in the case of either Hypothesis A or B are all pseudo-static strain rates, i.e., the upper bound of the range of strain-rates of interest is that strain-rate which does not result in inertial effects.

During primary consolidation, as the pore fluid drains and stresses become effective stress, particles move to final positions in a Poisson process. After some point (EOP), the number of particles that move to a final position is so small that the volume change is small and so, the pore fluid that needs to drain is so small that even if permeability is low, no excess pore pressures develop. Regardless of there being no pore pressures though, the same Poisson process that occurred during primary consolidation continues after EOP. Based on this, it is possible to show that the concept of a constant $C_{(\alpha e)}/C_c$ ratio (Mesri and Castro, 1987) has a rational basis.

In Chapter 6 on strain-rate effects in soil shear, DSSM based analyses showed that during shear, the static coefficient of friction dominates at small strains but that once particles mobilize, this static friction coefficient reduces to its dynamic value. Both the static and dynamic coefficients of friction are strain-rate dependent and responsible for the soil-structure viscosity. Temperature effects are very small and act only to change the nature of this dependence on strain-rate of the interparticle friction.

At present, it is unclear if for clays, the friction is between primarily particle material to particle material, or particle material to adsorbed water, or adsorbed water to adsorbed water or some combination thereof. The next section discusses this issue and suggests that the viscous effect is due to the adsorbed water layers surrounding particles. The basis of this claim is the remarkable finding by Watabe and Leroueil (2012) that absolute values of strain-rate dependent behavior measured in one-dimensional compression and triaxial tests appear to be approximately the same for clays across the world!

Chapter 5 also showed that variations with strain-rate in the stress-strain and void-ratio-strain curves during triaxial testing, i.e., viscous effects during triaxial testing, are small, due to the correspondingly small dependence of the friction coefficients on strain-rate and so, difficult to measure accurately. Consequently, the expectation is that viscous effects during one-dimensional loadings where strain-rates are small (primary consolidation and secondary consolidation in general), are likewise also small and difficult to measure.

---

respect to the other. Once we did this, the tank slapper went away forever, without adversely affecting the ability of the driver to steer (as the behavior-based solution of simply tacking on a damper to the steering did). Since then, it has always been key to me to determine the root cause that drives any phenomena–without that, one is not able to find the precise solution. Hence, an inbuilt understanding I have is this: behavior is secondary–root cause is primary and a root cause analysis reveals the underlying physical basis, i.e., the root cause of the behavior. Underlying process is primary because behavior is but a manifestation of the underlying process.

This chapter (Chapter 7) also details on this basis that the EOP based e vs. σ'_v relationship, though slightly strain-rate dependent, is unique for tests at a given strain-rate (CRS tests) or sequence of strain-rates (load controlled tests). Hypothesis B, discussed above, appears to be strictly valid, though practically, the effects of the viscosity of the soil-structure are small and can usually be ignored for most field situations.

The broad conclusions are that: a) current methods of calculating settlement based on hypothesis A are useful simply because viscous effects are negligibly small at the strain-rates, soil types, and layer thicknesses typical of most field situations and b) a valuable research program would be one targeted specifically at elucidating the role of the adsorbed layers in determining viscous effects.

SECONDARY CONSOLIDATION IS CONTINUED PARTICLE MOVEMENT

Secondary consolidation is a continuation of primary consolidation in terms of the physical particle movements except that the soil deformation is slow enough so that the soil is able to drain pore fluids without pore-pressures building up (see for example, Mesri and Vardhanabhuti, 2009). Chapter 3 (Joseph, 2012) showed that a Poisson process underlies particle movements during shear. Chapter 6 (Joseph, 2014a) generalized the dynamical system model that describes this Poisson process to include all shear-paths including that of shear along the K_o line, i.e., one-dimensional consolidation. This specialization of the general model to the case of K_o loading resulted in deriving the standard linear relationship between void-ratio and the log of the vertical effective stress, commonly observed.

The derivation specialized to K_o loading also showed that it is possible to write C_c (change in void-ratio per log cycle of effective vertical stress σ'_v) as $C_c = (1 + e_0)/\lambda$ where e_0 is the void ratio at the start of the test and λ the rate (per unit strain) at which particles move to their final position corresponding in the case of one-dimensional compression to the final effective vertical stress. (Note that λ is here different than in CSSM where it stands for the slope of the one-dimensional compression e vs. $\log \sigma'_v$ line.) From this it follows that the compression ratio CR, the change in strain over one log cycle of stress is $CR = C_c/(1 + e_0) = 1/\lambda$.

Mesri and Vardhanabuti (2009) provide evidence that in the case of granular materials with high permeability, particles have essentially finished adjusting to the effective stress in 10 to 20 seconds. Thereafter, deformations continue at a very small rate due to local, interparticle shear stresses at a few contacts, exceeding the contact's ability to support it over the long term, and consequently, small readjustments continue to take place at a rate such that strain decreases exponentially with time. In this chapter, the hypothesis is that when the vertical stress becomes fully effective, though most particles have moved to their final position, some particles remain unstable. Continued slippage of these quasi-stable particles continues to occur at random strains as local inter-particle contact shear stresses exceed locally available shear resistance, resulting again in a Poisson process with particles moving to new fully adjusted positions at random shear strains.

For the Poisson process that occurs in the case of one-dimensional consolidation, if the rate at which particles move per unit strain to their final position is λ and N_0 is the number of particles at some vertical effective stress σ'_v then the number of particles that remain to move to their final position at some subsequent vertical strain ε_v is $N_0 \exp(-\lambda\gamma)$ where $\gamma = \sqrt{2}/3 * \varepsilon_v$. Over an increase of one log cycle of stress, strain $\varepsilon_v = CR = C_c/(1 + e_0) = 1/\lambda$ and so $\gamma = \sqrt{2}/3\varepsilon_v = (\sqrt{2}/3)/\lambda$ or in other words, corresponding to a change in effective vertical stress of one log cycle, i.e., a stress of $10\sigma'_v$ the number of particles that remain to move to the final position is $N_0 \exp(-\lambda * (\sqrt{2}/3)/\lambda)$ or $N_0 \exp(-\sqrt{2}/3)$. In other words, the number of particles that moved to their final position is $N_0 - N_0 \exp(-\sqrt{2}/3)$ or $N_0(1 - \exp(-\sqrt{2}/3))$.

Corresponding to this new effective vertical stress, after one log cycle of time, the change in strain is $C_{(\alpha\varepsilon)}$. During this stage particles are moving to their final position at a rate at or less than λ and so, the lower bound of the number of particles that remain to move into their final position is $N_0 \exp((-\sqrt{2}/3)\exp(-\lambda(\sqrt{2}/3)c_{\alpha\varepsilon}))$ or the upper bound of the number of particles that moved in one log cycle of time is $N_0 \exp(-\sqrt{2}/3) - N_0 \exp(-\sqrt{2}/3\varepsilon - \lambda(\sqrt{2}/3)C_{\alpha\varepsilon})$ or $N_0 \exp(-\sqrt{2}/3)(1 - \exp(-\lambda(\sqrt{2}/3)C_{\alpha\varepsilon}))$.

In other words, the upper bound of the ratio of strain during one log cycle of time to the strain during one log cycle of effective vertical stress is $N_0 \exp(-\sqrt{2}/3)(1 - \exp(-\lambda(\sqrt{2}/3)C_{\alpha\varepsilon}))/N_0(1 - \exp(-\sqrt{2}/3))$ or $(1 - \exp(-\lambda(\sqrt{2}/3)C_{\alpha\varepsilon}))(\exp(\sqrt{2}/3) - 1)$, i.e., the largest value of $C_{\alpha\varepsilon}/CR$ or $C_{\alpha\varepsilon}/C_c$ is:

$$\frac{C_{\alpha\varepsilon}}{CR} = \frac{C_{\alpha\varepsilon}}{C_c} = \frac{1 - \exp(-\sqrt{2}/3\lambda C_{\alpha\varepsilon})}{\exp(\sqrt{2}/3) - 1} \approx 1.66(1 - \exp(-\sqrt{2}/3\lambda C_{\alpha\varepsilon})) \quad (1)$$

Given that this is during secondary consolidation, when deformation proceeds at a slow rate so that the drainage ability of the material exceeds the need to drain pore fluid and that consequently, pore-pressure does not develop, $C_{(\alpha\varepsilon)}$ is very small and $\sqrt{2}/3\lambda C_{\alpha\varepsilon}$ is near zero. This means that $\exp(-\sqrt{2}/3\lambda C_{\alpha\varepsilon})$ is near one or the numerator of Equation 1 is near zero, and so is insensitive to the specific value of $C_{(\alpha\varepsilon)}\lambda$ because this value is small.

For example, Mesri and Castro (1987) show values of $C_{(\alpha\varepsilon)}/C_c$ for 43 tests on Berthierville clay (a Canadian quick clay, initial water content between 56–61%, a LL of 46% and a PL of 24%; USCS classification: CH). The average value from direct measurements of $C_{(\alpha\varepsilon)}/C_c$ was 0.045. From their data, the value of $C_{(\alpha\varepsilon)}\lambda$ used in Equation 1 averaged 0.102, a small value that results in an average calculated value for $C_{(\alpha\varepsilon)}/C_c$ from Equation 1 of 0.078. Likewise, Mesri and Castro (1987) also provide data from 36 tests of a tar sand and 24 tests of a shale $C_{(\alpha\varepsilon)}/C_c$. From their data, the value of $C_{(\alpha\varepsilon)}\lambda$ averaged 0.044 and 0.033 respectively, again, small values that in turn resulted in values of $C_{(\alpha\varepsilon)}/C_c$ calculated using Equation 1 of 0.034 and 0.025. These compare well with their reported values from direct measurements of $C_{(\alpha\varepsilon)}/C_c$ of 0.035 and 0.029 respectively.

In short, the calculated values from Equation 1 are close to the actual measured values, and more importantly, the root cause for the insensitivity of $C_{(\alpha\varepsilon)}/C_c$ is the form of Equation 1 that renders it insensitive to small values of $\sqrt{2}/3\lambda C_{(\alpha\varepsilon)}$. This is the only non-constant term in this equation, appearing as a negative exponent of the Euler number. This value, $\sqrt{2}/3\lambda C_{(\alpha\varepsilon)}$, is the number of particles that move to the final position during one log cycle of time during secondary consolidation, a value that *de facto* is small so that the soil deformation can occur slowly enough to permit free drainage without a buildup of excess pore-pressure.

VISCOUS EFFECTS

Dynamical systems soil-mechanics holds that a Poisson process in which friction at interparticle contact areas play a fundamental role drives soil deformation. Chapter 5 analyzed strain-rate effects in soil shear in the context of dynamical systems theory. One conclusion reached was that the variation in the steady-state strengths with strain-rate was similar to the variation of the maximum shear strength, even though the soil structure obtaining at the steady-state is the flow structure, and different than the soil structure at maximum strength. In other words, strain-rate effects appear to be independent of structure. As always, it is important to note that the strain-rates under consideration are all pseudo-static, i.e., bounded at the upper limit by the maximum strain-rate that does not cause inertial effects.

Based on these data, Joseph and Graham-Eagle (2013) suggested that interparticle static and dynamic frictional coefficients control strain-rate effects. This interparticle friction could be particle material to particle material or between adsorbed water layers, or some combination of the two. Chapter 6 showed that the same Poisson process that applies for shear to the steady-state also applies for one-dimensional compression, and that consequently, interparticle friction also plays a key role in one-dimensional compression. Given the lateral constraint to movement of soil particles in one-dimensional consolidation, the strain-rate effects observed for one-dimensional tests will be an upper bound, as compared with triaxial tests that allow for lateral movement of particles.

Sheahan (1991) conducted 28 compression and 10 extension triaxial tests at various strain rates on uncemented, resedimented Boston Blue Clay—a glacial outwash of illitic CL clay deposited in a marine environment. He one dimensionally (K_0) consolidated samples to different Overconsolidation ratios (OCRs), then sheared them undrained at various constant strain-rates in monotonic compression and extension.

In general, peak shear stresses (calculated as half the difference between the vertical and horizontal effective stresses and normalized by the initial vertical consolidation stress) increased less than 4% percent per order of magnitude of strain-rate, while normalized effective vertical and horizontal stresses at the peak shear-stress varied an average of about 8% and 4% respectively per order of magnitude of strain-rate, for strain-rates between 0.05 and 50%. Figures 7.1a and 7.1b, show respectively the variation of the normalized vertical and horizontal stresses with strain-rate at different OCRs for the Sheahan data. The correlation coefficients, especially for the normalized horizontal stress data are low.

Sheahan and Watters (1997) conducted CRS tests on Boston Blue Clay at strain-rates of between 1 and 3% per hour (between 2.8 EE-04 s^{-1} and 8.4 EE-04 s^{-1}) together with incremental loading tests. Given that for load controlled tests, Mesri and Feng (1986) show a strain-rate at EOP of 2 EE-05 s^{-1}, the strain-rates used by Sheahan and Watters are well above those typical of secondary consolidation. Sheahan and Watters found the Compression Ratio (CR) to be unchanged with strain-rate. They show an increase in strain of about 10% for CRS tests at a strain-rate of 1%/hour (2.8 EE-04 s^{-1}) over tests conducted by incremental loading test (approximated at 2 EE-07 s^{-1}), i.e., a change of about 3% per log cycle of strain-rate. Likewise, the tests at a strain-rate of 3% per hour (8.3 EE-04 s^{-1}) showed a similar change per log cycle of strain-rate as compared with the incremental loading tests.

For Sheahan's triaxial tests on Boston Blue Clay, the vertical and horizontal stresses increased by 3.7% and 2.3% respectively per log-cycle of strain-rate, which compares well with the increase of 4% per log-cycle of strain-rates Sheahan and Watters (1997) saw for their one-dimensional compression tests on Boston Blue Clay. They suggested that compression behavior for remolded clays did not have the same degree of strain-rate dependence as intact sedimented clays. Nonetheless, the data are sparse and presently not enough to decide if these values hold in shear for a wide range of clays, as was the case for the one-dimensional compression data for worldwide clays as reported by Watabe et al. (2012).

Diaz-Rodrigues et al. (2009) give a concise summary of strain-rate effects seen for clayey soils in shear. Their review shows that undrained strength triaxial tests increases between 5–15% per log cycle of strain-rate. Their review also points out the contradictory results that research programs investigating strain-rate effects report—for example, that Zhu et al. (1999) and Zhu and

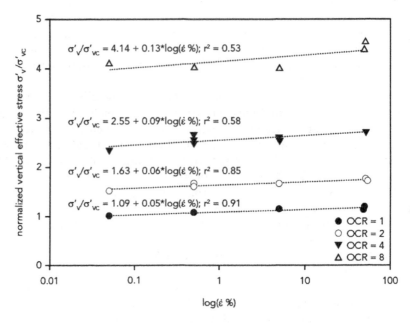

Figure 7.1a Variation of normalized vertical effective stress with strain-rate from Sheahan (1991) data.

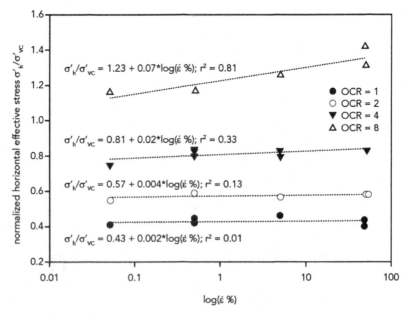

Figure 7.1b Variation of normalized horizontal effective stress with strain-rate from Sheahan (1991) data.

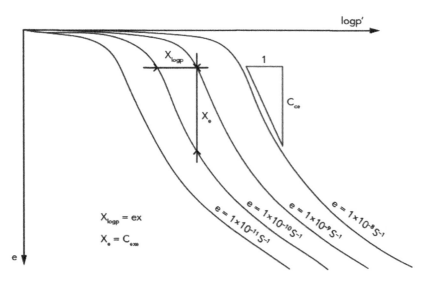

Figure 7.2 Illustration of geometrical relationship between the parameters (Figure 19 of Watabe *et al.*, 2012).

Yin (2000) report that strain-rate effects are higher in K_o consolidated samples versus isotropically consolidated samples whereas Graham *et al.* (1983) saw no significant effect for anisotropic consolidation.

A striking feature of the test programs that study rate-effects is that they appear to simply observe the results of varying strain-rates, but have no underlying root cause hypothesis that they then seek to falsify i.e., remain at the superficial level of phenomena instead of probing to the root cause, the physical basis of the phenomena. A useful test program would be one specifically organized for purposes of falsifying a root cause hypothesis with a physical basis to the effect that adsorbed layers of water acting at contacts between particles control rate effects, regardless of whether shear deformation is in one-dimensional compression or to failure and beyond in triaxial shear. Since the only difference between shear deformation in one-dimensional compression and triaxial shear to failure is that the former has a stress path that follows a K line, the expectation is that the same physical mechanisms underlie both types of deformation under loading. If one examines Figure 7.2 (an idealization of one-dimensional compression tests run at different strain-rates reproduced from Watabe *et al.*, 2013) but with the void-ratio axis horizontal, one sees that the effect of strain-rate in one-dimensional compression is qualitatively not unlike that seen in data at different strain-rates from triaxial tests.

HYPOTHESIS B IS VALID, BUT VISCOUS EFFECTS ARE SMALL

There has been extensive debate on which of Hypotheses A and B is correct (see for example, Ladd *et al.*, 1977 and Jamiolkowski *et al.*, 1985). The discussion assumes that the strain-rates of interest are all pseudo-static.

Figure 7.3, shows compression curves obtained for a layer of clay under Kansai International Airport constructed on an artificial island in the middle of Osaka Bay in Japan. Also, shown in the figure are the isotache curves of Hypothesis B corresponding to one-dimensional compression at different strain-rates for this same clay averaged from data from five tests, obtained from 24-hour incremental loading in an oedometer.

The reference strain-rate is $1EE-07\,s^{-1}$ a rate chosen to normalize the strain-rate data and which approximates strain-rate seen in long term 24-hour load increment consolidation tests in the laboratory. This 24-hour based curve is different from the EOP curve as it includes any secondary consolidation that occurred if consolidation completed in less than 24 hours. Given that for load controlled tests, Mesri and Feng (1986) show a strain-rate at EOP of $2EE-05\,s^{-1}$, the isotaches shown in Figure 7.2, all correspond to the secondary consolidation stage and give no information on strain-rates during primary consolidation.

The in-situ field data plotted in Figure 7.3 show that as the rate of compression decreased, the field curves traversed across the isotaches. The data

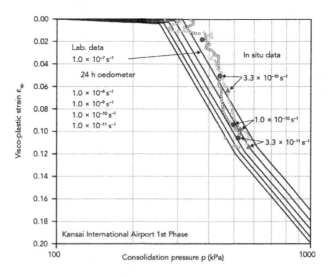

Figure 7.3 Compression curves for sample under Kansai International Airport with isotachs (Figure 10 of Watabe and Leroueil, 2012).

also shows that for this clay, the vertical consolidation stress corresponding to a strain-rate of 1EE-11 s^{-1} was 400 kPa and at a strain-rate of 1EE-07 s^{-1} was 500 kPa, i.e., a difference of about 6% per order of magnitude of change in strain-rate.

Watabe *et al.* (2008, 2012) and Watabe and Leroueil (2012) show the relationship between compression curves and strain-rate for clays from all over the world. Figure 7.4 shows this relationship. The striking feature of this curve is that it applies to a wide range of inorganic clays, regardless of variations in plasticity, mineralogy, fabric, microstructure, overconsolidation, etc. Included even is Mexico City clay, a clay noted for its exceptional characteristics as compared with other soils, and with a water content of near 400%. The data also includes soils from Italy, Japan, Canada, and Sweden. This is a remarkable finding, and if it holds up, is truly worthy of commendation!

Figure 7.4 suggests that the strain-rate dependence in one-dimensional compression is the same for clays all over the world and is insensitive to the actual particle material. This seems to indicate that the adsorbed water layers surrounding particles play a fundamental role in strain-rate dependence, and factors such as actual soil particle material, fabric, structure, etc. only play

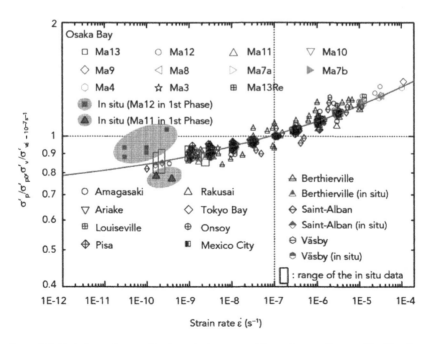

Figure 7.4 Variation of normalized vertical stress with strain-rate for world-wide clays (Figure 7.7 of Watabe and Leroueil, 2012).

secondary roles. While these latter factors may have a strong role in determining the soil's other engineering properties, the data from the world-wide clays indicate that so far as the direct dependence of frictional behavior on strain-rate is concerned, such considerations are secondary. Such a hypothesis if true, would significantly simplify research in soil strain-rate behavior, locating it intrinsically in the adsorbed water layers. Recall also that about 70 years ago, Terzaghi and Peck (1948) suggested that the adsorbed water layers played a fundamental role in determining clay behavior.

Examination of the curve in Figure 7.4 shows that between the strain-rates of 1EE-11 and 1EE-07 considered for the Kansai International Airport clay, the normalized consolidation pressure changed by about 25%, i.e., again, a change of about 6% per order of magnitude change in strain-rate. At higher strain-rates, for example between 1EE-07 and 1EE-04, the change was higher, approximately 10%.

Watabe et al. (2013) show that if the isotaches follow the idealization shown in Figure 7.2 then, $C_{(\alpha\varepsilon)}/C_c = (\Delta\log p_c')/(\Delta\log\varepsilon) = \alpha$ where α is a constant. Given that this constant ratio is due to the random movement of particles from their quasi-stable state to their final state in a Poisson process, it is independent of any substantial viscous contribution, i.e., independent of either Hypothesis A nor Hypothesis B.

Because the same mechanism of deformation that occurs during primary compression occurs during secondary consolidation except at a rate slow enough to allow free drainage, the same dependence on strain-rate at interparticle contacts that holds during primary compression should continue to hold during secondary consolidation. However, in the case of secondary consolidation, this strain-rate is very small. Nonetheless, viscous effects, very small though they may be, should theoretically exist during secondary consolidation.

Watabe et al. (2008) ran 24-hour incremental loading tests in a special oedometer to evaluate scale effects in long term consolidation. They took care to account for frictional losses arising at the interface between the sample and the oedometer ring. Their conclusions lend support to Hypothesis B, i.e., viscous effects occurring both during and after primary consolidation. They found that the law of H^2 is essentially (i.e., not exactly) valid for laboratory specimens between 20 to 200 mm. For clay samples without a developed structure, strain variation followed Hypothesis A but that for samples with a developed structure, it followed Hypothesis B though even in this case, at thicknesses greater than 50 mm, strain variations followed Hypothesis A. They found that a structured clay will collapse at a higher strain-rate than a non-structured clay and that larger viscous effects would accompany these higher strain-rates even during primary consolidation, i.e., in accordance with Hypothesis B.

In a discussion of Watabe et al. (2008), Mesri (2009, pp. 823), quotes Mesri (2003) as follows: "As soon as primary consolidation begins, both

$(\delta e)/(\delta\sigma'_v)_t$ and $((\delta e)/(\delta t))_{(\sigma'_v)}$ contribute to compression; however, only $((\delta e)/(\delta t))_{(\sigma'_v)}$ contributes to compression during secondary consolidation when $(d\sigma'_v)/(dt) = 0$." (Note: t is time.) However, it isn't clear from Mesri's quote what would prevent viscous effects from occurring during secondary consolidation, added to continued Poisson based particle movement.[3]

Mesri, in his various writings quotes the uniqueness of the EOP curve under static loading (see for example, Mesri and Choi, 1985), i.e. that this curve is unique, and independent of the duration of primary consolidation, i.e. of layer thickness. As evidence of the uniqueness of the EOP curve and its independence on the duration of primary consolidation he gives data from Aboshi (1973). In this paper, Aboshi has measured values of EOP strain increments of clay layers of different thickness. The data appear to show that although for a given vertical stress the layers of different thickness took different times to finish primary consolidation, the strain increment at the end of primary consolidation from the start of consolidation was the same.

Given that the smallest sample thickness for Aboshi's tests quoted by Mesri is 60 mm, this matches the behavior seen by Watabe *et al.* (2008) where for sample thicknesses greater than 50 mm, strain variations followed Hypothesis A. Note that others have interpreted the Aboshi data (see for example, Yasuhara, 1982; Leroueil *et al.*, 1985; Imai and Tang, 1992; Hawlader *et al.*, 2003; Li *et al.*, 2004; Watabe *et al.*, 2009). Oka *et al.* (1986) and Oka (2005) interpreted Aboshi's data to be between hypotheses A and B (which they called Hypothesis C). In the closure to their paper, Watabe *et al.* (2009) reiterate that all consolidation behavior follows hypothesis A when the layers are thick regardless of whether the clay has a well-developed structure or not, and that the key factor is the threshold between thick and thin clay layers, i.e., the drainage path length.

Small drainage paths, and structured clays, both result in higher strain-rates for a given loading suggesting a stronger role for viscous effects during consolidation with measurable effects of particle level viscous behavior, i.e., would support Hypothesis B. Larger drainage paths and unstructured clays that do not collapse will show smaller strain-rates under loading and in smaller, hard to measure viscous effects during consolidation and so would appear to support Hypothesis A.

In short, particle level viscous effects occur always, both during and after primary consolidation, i.e., Hypotheses B is valid while Hypothesis A is not.

---

[3] Here we see the peril of approaching a problem merely at the phenomenal level and without any underlying theory based on the physical root cause of the phenomena being observed. Viscous effects during creep are necessarily very small given the small strain-rates during creep, and so very hard to measure, i.e., easy to miss if one is not aware it must be present. This awareness may in turn require more sophisticated measurements to accurately profile the phenomena in question.

Ramaswamy (2014), pithily summed up this chapter: viscous effects occur all the time; depending on strain-rate, they may or may not be measurable.

At higher strain-rates, viscous effects are measurable, while at lower strain-rates, they are small and hard to distinguish from changes due to particle rearrangement. Consequently, though Hypothesis B is valid, nonetheless, strain-rates in the field being low, design calculations based on Hypothesis A should suffice.

The exception is for thick deposits of highly compressible soils where strain-rate effects may be significant (Grimstad, 2014). For such cases it would be prudent to also calculate settlements using the Isotache Method such as that proposed by Watabe and Leroueil (2012). The isotache process needs very high quality soil sampling in the field for to give meaningful results as the pre-consolidation pressures measured strongly influence the results of design calculations done using either Hypothesis.

As per Hypothesis B, the EOP based curve is strain-rate dependent, a dependence readily seen in constant rate of strain (CRS) tests, see for example Degago et $al.$ (2011). However, for static load controlled tests, given that each stress increment results in its own unique strain-rates through the consolidation process, the EOP based e vs. σ'_v relationship would indeed also be unique. This contrasts with the case for CRS tests where the e vs. σ'_v relationship is also unique but only for the given strain-rate used for the test.

CONCLUSION

Driving the viscous and secondary consolidation phenomena seen in one-dimensional compression down to underlying physical mechanisms furthers understanding of these phenomena and how they determine soil behavior in the laboratory and field. Dynamical systems soil-mechanics suggests that viscosity effects and secondary consolidation each have a separate physical basis.

Viscous behavior occurs both during and after primary consolidation in accordance with Hypothesis B and is due to strain-rate dependence of the coefficients of friction at interparticle contacts as they slide against each other in simple friction. The similarity of strain-rate effects across different soils and stress-paths suggests that the physical mechanisms underlying them are located in the adsorbed water layers surrounding the soil particles. The EOP curve does depend on strain-rate. Consequently, for load-controlled consolidation, where each load has its own unique strain-rate, the resulting EOP based e vs. σ'_v relationship must also be unique.

Secondary consolidation is the continued deformation of the soil structure after consolidation due to small numbers of particles moving at random shear strains, in a Poisson process, to new final positions. The near constant $C_{(\alpha e)}/C_c$ ratio is due to the form of the equation derived from the underlying

Poisson process and to the fact that very few particles move to new positions during secondary consolidation. It has little to do with viscous effects given that predictions of $C_{(\alpha e)}/C_c$ based on this model match the empirical data. The fact that $C_{(\alpha e)}/C_c$ is approximately constant has no direct bearing on either Hypothesis A or B. Rather it is driven by the underlying Poisson process (and not strain-rate dependent friction).

Given that strain-rates involved in field situations are pseudo-static and very small, and that clay layers are usually not too thick, viscous effects in field problems will be small during either primary or secondary consolidation. Consequently, the current design practice of using Hypothesis A as the basis of calculating both short and long term compression settlements seems reasonable, except for unusual cases such as very thick deposits of highly compressible clays (for ex. the clay layers under the artificial island built for the Kansai International Airport in Osaka Bay, Japan).

Again, the power of a simple theory is that it lays bare the underlying physical mechanisms and consequently can explain behaviors (in this case, C_α/C_c being constant and the EOP curve from static loading being unique) hitherto taken as granted or as in the case of Hypothesis A vs. Hypothesis B, hitherto unresolved.

At this point in the book, strictly speaking, some would say the soil-mechanics portion is over! The next chapter is about using the DSSM model in finite element analysis and this should be considered geotechnical engineering and not soil mechanics (Iglesia, 2013). This is because the way finite element analysis is done in geotechnical engineering does not meet the requirement of a science, i.e., the requirement that a result be falsifiable by empirical evidence. Publications in journals that report the results of finite element analyses, regardless of the model used, show results that are (surprisingly) always approximately correct. This is because results are typically what Lambe (1973) called Class C predictions–the worst kind of prediction, i.e., predictions made with the result already known.

Before reading the next chapter though, closely examine the image on the front cover of this book. Introspect and come up with at least two reasons as to how the DSSM model is able to correctly account for the irregular shapes and variously sized (finite) soil particles, without explicitly analyzing the shapes and finite sizes of the particles.

Chapter 8

Finite element analysis using DSSM

This chapter shows how to incorporate the DSSM model into finite element analysis (FEA). The use of FEA dates from at least the 1950's with its application to membrane and plate bending problems in the aeronautical industry—Clough and Wilson (1999) cover the history of its development. A milestone in FEA development was the publication in 1960 of its use in determining stress-strain distributions in continuous plane-strain concrete structures (Clough, 1960). Clough (1962) describes his use of a finite element analysis using a non-linear model to analyze a concrete gravity dam for the US Army Corps of Engineers.

The method then attracted the attention of O. C. Zienkiewicz who further developed it significantly. Clough, as a visiting professor taught the method at Cambridge University and on returning to the USA, he applied the method to various engineering problems engendered by the Cold War, primarily problems in the aerospace industry and underground nuclear shelters. Use of the method gradually spread from the aeronautical industry and the analysis of concrete structures to the analysis of problems in other areas including geotechnical engineering.

FEA is but a way of numerically solving differential equations that describe the variation of some quantity over some spatial area or volume. The basis of the finite element method consists of modeling the problem in terms of its geometry using an assembly of small regions called "finite elements" made up of nodes joined by lines that form the boundaries of the elements.

In soil mechanics problems, either stress or displacement is the primary variable and rules established as to how this variable must distribute over the finite element model to satisfy equilibrium conditions. A "constitutive model" based on the nature of the material being analyzed converts node displacements to nodal forces. In general, most finite element analyses of geotechnical problems use metals based elasto-plastic models—usually some formulation of the CSSM model originally proposed by Roscoe's research group at Cambridge University (Roscoe and Schofield, 1963 and Roscoe and Burland, 1968).

SOILS ARE NOT METALS[1]

This modeling of soils as if they are metals is a deeply flawed approach that found much use over the years, simply because there was no good alternative.

---

[1]In 1980, one of the first questions I asked my soil mechanics professor S. V. Ramaswamy was why were soils modeled as metals. I had just turned 20, and by this time, my brother and I had been tuning two-stroke motorcycles for racing, for several years. At that time in India, motorized metal grinders were not cheap and so we had to use ordinary metal hand files to raise or lower the two-stroke intake, transfer and exhaust ports. The experience of grinding cast iron manually was for me, simply put, a huge shock. Only when I took 8 hours to lower the exhaust port by 3 mm did I realize how hard a metal (then also, a relatively soft metal like cast iron) really was. Only when I saw the fine iron powder, in which it was impossible to discern any different shapes of the iron powder particles without a microscope, did I realize what an atom must be. Hence, when told in undergraduate class that soil mechanics modeled soils as metals, I was instinctively and at once taken aback–the idea struck me as simply absurd–hence my question. It was Professor Ramaswamy, the person who sparked my interest in soil mechanics when I was an undergraduate, who told me that it was possibly because of his training as a mechanical engineer, as well as the lack of any alternate theory, that made Roscoe amenable to suggestions of modeling soils as metals. Today, almost thirty-five years later, I realize the importance of my experience filing metal with a hand file for eight hours. This is exactly why I admire Nietzche's saying: The doer alone learnth. As with metals, so also with soils! Hands on contact is essential to obtain a physical feel for the object of study and to truly understand! Today, sadly, most academics lack this "physical," hand on training, and hence are too quick to accept statements like "soils are really metals in disguise." Steve Poulos firmly believed in hands on contact with soils to gain this knowledge. Likewise, he told me, did his boss, Arthur Casagrande who made Steve spend a lot of time in the soils laboratory. And as it happened with Steve, so also it happened with me. A book that captures this view was a New York Times non-fiction best-seller. I consider it essential reading if you want to become very good at your profession, whatever it may be. The book is Shopcraft as Soulcraft: An Inquiry into the Value of Work (Crawford, 2010). I think this book is essential reading if you want to become very good in soil mechanics as well! I now realize that what Roscoe and the "metal" people knew but which I did not then, was that at very high stresses, metals indeed behave like "modeling clay." However, what I intuitively realized then and which Roscoe and his "metal" people did not seem to (or at least, to this date are not able to realize in their model) is that while metals may constitute of chemical molecules, i.e., effectively, isotropic point particles, real soils are not such. Rather, particles with very anisotropic shapes constitute real soils. This seemingly trivial difference is the heart of the matter, the very core of it. And this is the basis of my deliberate choice of the image on the cover of this book–a photograph showing soil particles of finite size and irregular shapes. If you look closely in the upper left of the photograph in the black and white part of the image, there is but one somewhat spherical particle present that perhaps, could be approximated as isotropic! Models that do not account for this core anisotropy and finite size of soil grains are bound to fail, just as current "metal" models of soils have failed. This is because finite size anisotropic grains create complex physical structure that resembles a "house of cards." Anisotropic grains also have irregular shapes by definition. And it is this structure and grain shape that controls behavior. Metal-based theories of soil do not

Its origins trace to the Drucker-Prager criterion first proposed in 1952 by the two mathematicians Daniel C. Drucker and William Prager in an eight-page note in a journal of mathematics (see Drucker and Prager, 1952). This failure criterion was for materials idealized as having no structure and behaving plastically during deformation. In their note, Drucker and Prager also showed how to use their theory to calculate the critical height of a vertical bank of soil using plane and log spiral failure surfaces.

Roscoe, who trained as a mechanical engineer before switching to soil mechanics and was then the head of the soil mechanics department at Cambridge University, enthusiastically embraced the Drucker and Prager model and together with Wroth and Schofield, launched the use elasto-plastic based analyses of soils.

Modeling soils as metals is a fundamentally flawed, failed, and broken approach–simply put, soils are not metals. Implicit in assuming metal like behavior is that soil particles are isotropic point particles. They are not— they have finite size with particle level anisotropy, that is not accounted for in elasto-plastic theories, but which control deformation. Note: see the image of a real soil on the cover of this book, a photograph deliberately chosen to show that real soils are anisotropic and have finite size.

The question then arises–why has the soils community stuck with such models for so long? One reason is that that, yes, crudely, very very crudely, a fine grained, homogenous material, lacking in structure (think for example of modeling clay), approximately resembles a yielding, metal. Perhaps Roscoe naturally analogized soils with metals because he was a Mechanical engineer by training. CSSM was berthed in the 1950's when most triaxial tests were on fine grained fat clays, remolded and reconsolidated isotropically, and tested to a few percent of axial strain. Such tests generate simple stress-strain-volume curves and scarcely exhibited strain-softening; indeed, a metals theory of plasticity crudely approximates their behavior.

An assumption implicit in metals based elasto-plastic soil models is that soil particles are isotropic point particles. Because most real soils, with the exception of very homogeneous fine-grained clays and silts, have a significant amount of particles greater than clay size, the approximation fails abysmally. As their presence increases, the effects of these anisotropic soil grain properties begin to play a progressively more significant role and consequently metals based plasticity models fit empirical test data poorly, particularly data

---

capture this card like structure or the irregular shape properties resulting from the core property of grains–that they are anisotropic at the particle level. Consequently, such metal plasticity based models are fundamentally broken at their very core. The center of the theory does not hold, and so things fall apart! Attempts to model directly such card like structure will result in extremely complicated mathematics. DSSM on the other hand does not need to model this structure explicitly because the friction based Poisson process incorporates the net effect of this structure at the grain level.

on volume change or pore pressures. Hence, the most that metals based models can do is give a crude approximation of true soil behavior and then too, only for fine grained soils such as CH clays with very few particles greater than clay size.

Further, metals based elasto-plastic soils models often are naive in terms of fundamental underlying theory–most assume for example that hydrostatic stress does not cause shear deformation. This is patently false–a soil structure is analogous to a house of cards and the application of pure hydrostatic stress to a real soil causes significant shear deformation. Andrew Schofield (see Schofield, 2005) tried his best to put soil samples into pure compression in centrifuge tests and found he was not able to for the reason described above–soils are not point particles, i.e., they are not isotropic–simple as that. Consequently, they form structures whose grains, when subject to pure hydrostatic compression, slip, causing shear deformations. The DSSM model, explicitly captures this interdependence of deformations caused by shear and hydrostatic forces.

In short, the belief that soils are "really metals" qualifies as a falsified hypothesis because a wide range of readily available empirical data readily falsifies it. Remember from the Introduction to this book what Feynman said! "You may have the most beautiful and elegant theory in the world, but if the model that results from it does not fit the data–then your theory is simply wrong." It is easy to falsify an elasto-plastic soil models using real test data from any soil with a wide distribution of grain sizes and which exhibits strain-softening behavior.

By contrast, the DSSM model does not need these naive (soils are point particles) and dangerous (separate hydrostatic and shear stresses and deformations) assumptions. Recall from Chapter 4 that DSSM does not treat soils as isotropic point particles and that the DSSM model (inspect equations 1 in this chapter) correctly predicts shear stresses and deformations when applying pure hydrostatic stress and vice-versa. In short, DSSM models real soils, which have particles with anisotropic shapes and properties and which are larger than clay size.

IS FEA SCIENTIFIC?

The next question is this–does FEA count as mechanics or engineering? FEA at its core, i.e., in terms of its theory and formulation, counts, as scientific, if every part of it, particularly the constitutive model it uses as well as the results, pass the criteria of what it takes to count as scientific–testable, repeatable, and falsifiable. Hence, if the constitutive model remains unfalsified outside of the "standard machinery of FEA," (more on this "standard machinery" next), then yes, FEA can be, in theory, scientifically (i.e., falsifiably) applied to a "fully known problem." This "fully known problem" is likely to be a problem of very limited scope, one where it is possible to know with a good degree of certainty, model inputs, properties, and outputs.

Very limited scope means, for example, triaxial tests or consolidation tests on homogeneous samples, where it is possible to calculate using the constitutive model outside of FEA, behavior in terms of their stress-strain and volume/pore-pressure changes, and compare these with the empirical evidence, i.e., the actual test data. This is the approach taken in this chapter. Yet, at the end the fact remains that in practice, there is no good empirical data with which to compare all the results, even for the case of the simple triaxial test. For example, there is no way presently to measure the stresses and strains inside the triaxial test specimen. Hence, though FEA programs (as in this chapter) can readily produce "pretty pictures" showing internal loads and displacements, it is not practically possible to determine if these internal loads and displacements are accurate or not.

In short, though theoretically it may be possible to falsify or fully confirm the result of FEA analyses predictions, in practice it usually isn't possible to do this even for simple problems, i.e., applications of FEA to soil problems currently does not qualify as "science."

FEA FRAMEWORK

Broadly speaking, FEA consists of a framework consisting of three "standard" parts plus a fourth non-standard part. The three standard parts, referred to in this chapter as "the standard or generic machinery of FEA" are: i) a relatively small number of standard algorithms that consists of routines that model the geometry of the problem, ii) a relatively small number of standard algorithms from linear algebra that deal with the manipulation of matrices, iii) a set of numerical procedures that typically vary with the domain–for example, in soil mechanics, a common numerical procedure used to study pore-pressure dissipation with time is the generic Crank-Nicholson method.

The Crank-Nicholson method is a very powerful one that applies to many kinds of FEA problems of which diffusion problems are only one kind.[2] With time and practice, these three parts will become second nature–they are usually always the same, repeated in problem after problem. In most cases, it is safe to treat the routines that implement them as "black boxes" where data goes in, and the output simply used. This is why this chapter refers to these three parts as "standard FEA machinery."

---

[2] Yes, I am afraid that geotechnical engineering is not the center of the FEA world, a lesson that I learned with much surprise!

It is only in very unusual conditions that one needs to figure out how these routines really work. For example, the work described in this chapter–that of incorporating a brand new constitutive model into a FEA is by itself a rare and unusual task simply because it is not often that one must incorporate a brand-new model into a FEA. Yet, rare and unique though this task may be, as this chapter shows, it turns out that only a very few of the routines that count as part of "standard FEA machinery" need modification.

The fourth part is the constitutive model that one chooses for the analysis. The challenge is to fit it in with the other three parts, i.e., to fit it in with the standard machinery of FEA. Consequently, this chapter has only a very little discussion of the modeling of geometry and only a little bit more on the numerical methods. In short, the focus of this chapter is the incorporation of the actual constitutive model–in this case the DSSM model–into FEA.

Problems such as the dissipation of pore-pressures using standard methods such as Crank-Nicholson are "left to the reader" as they (in)famously say! Note that for the DSSM model properties vary with strain and so must be recalculated each time the strain changes using the equations for the soil type and stress path, i.e., the equations corresponding to the regression equations shown in Figure 3.2 of Chapter 3. This is a standard approach and is described in "Programming the Finite Element Method" (PFEM) by Smith *et al.* (2014).

To incorporate the DSSM model into FEA, there are three main tasks. These are 1) formulating the force equilibrium equation that governs in terms of the DSSM model, first using continuous functions, then, discretizing these functions, 2) expressing this discretized force equilibrium equation in terms of a matrix formulation that is amenable to numerical analysis, and 3) implementing the matrix formulation and resolving any numerical issues that may crop up.[3]

A warning about the next section: it assumes knowledge of the theory behind Finite Element Analysis (FEA) and of how to write finite element

---

[3]The first step took me about two months. The second step, though in hindsight trivially obvious and simple (in doing this work, a lesson I learned the hard way, is that in hindsight everything seems "trivially obvious and simple"), took me a long time–six months–one month to re-familiarize myself with FORTRAN and the development tools I needed, plus an astonishing five months to figure out how to formulate a square element stiffness matrix in terms of the nodal displacements. The final step–implementation and resolution of numerical issues took me a month. (A friend of mine has this thumb rule: any substantial human task takes a minimum of nine months!) Granted, I was doing all this "on the side" of my real job, but truth be told, that is no real excuse because most of the time I was stuck and could only stare into the void, blocked, and waiting for the answers to present themselves. I took great inspiration from Newton's words: "I keep the subject constantly before me, and wait 'till the first dawning's open slowly, by little and little, into a full and clear light." If I knew then what I know now, the work described in this chapter should have taken me not more than a weekend to do!

programs. This is a much deeper level of knowledge than that needed simply to run a finite element program that somebody else wrote, and which only needs construction of its associated data file. A good book to learn how to code finite element programs is the PFEM book mentioned above.[4] *Until that time though, skip this next section safely; go to the section after the next one. That section describes an example implementation–the FEA of an undrained triaxial test, and its results.*

DSSM BASED FORCE-EQUILIBRIUM EQUATION

Central to the FEA of geotechnical problems involving stresses are the force equilibrium equations for the nodes that make up the finite element mesh. This section describes the development of these equations in terms of the DSSM model. With the exception of the constitutive model for which, instead of using some variant of the elasto-plastic model, we use instead the DSSM model, the rest of the FEA is "generic." It uses the same standard machinery described in the previous section for specifying the geometry, the matrix manipulations, and the numerical methods, etc.

To recap, DSSM model is as follows:

$$\frac{dq}{d\gamma} = J_q[\bar{p}\tan\alpha_q \exp(-\lambda_p\gamma) - q\exp(-\lambda_q\gamma)] \tag{1a}$$

$$\frac{d\bar{p}}{d\gamma} = J_p[\bar{p}\tan\alpha_p \exp(-\lambda_p\gamma) - q\exp(-\lambda_q\gamma)] \tag{1b}$$

$$\frac{-de}{d\gamma} = -J_e[\bar{p}\tan\alpha_e \exp(-\lambda_p\gamma) - q\exp(-\lambda_q\gamma)] \tag{1c}$$

where: $e =$ void-ratio, $q =$ shear stress, $\bar{p} =$ effective normal/confining stress, $J_q, J_p =$ initial non-dimensional inter-particle contact area of the load

---

[4]In fact should you read the PFEM book you will see that what I have described above is nothing but their "building blocks strategy," a strategy that has served the book well over the last thirty years it has been in print. The example I analyze in this chapter, is their example 6.11 (5th edition of the book), except for the important fact that while they use the Mohr-Coulomb failure criterion in most of the examples in Chapter 6 of their book, I use instead the DSSM constitutive model. That being said, the Mohr-Coulomb failure criterion and also the von Mises yield criterion are not accurate in that they implicitly assume that soils are constituted of isotropic point particles and their implementation in FEA programs arbitrarily separates the effects of compression from shear. Because of these implicit assumptions such models can only model soil behavior very crudely and their inclusion in this book is likely for their pedagogic value. Do not use either the Mohr-Coulomb or the von Mises criteria (or any other criteria that similarly treat soils implicitly as point particles) for real world geotechnical problems.

bearing particles that are not in the final structure, per unit area of the sample, corresponding to q and $\bar{p}$ respectively, J_e = void-ratio due to load bearing particles not being in the final structure, per unit stress, per unit area, α_q, α_p, α_e = secant friction angles corresponding to q, $\bar{p}$, e respectively, γ = shear strain, and λ_q, λ_p = rate of movement of particles into the final structure, corresponding to q and $\bar{p}$ respectively, per unit of strain.

The goal in this section is to incorporate these equations into a FEA. The key challenge posed by the dynamical systems model is that the older elasto-plastic models originating with the Cam clay models use a constitutive relationship that directly connects principle strains to stresses, whereas the DSSM model relates increments in shear strain to the increment of shear and confining stresses in terms of current shear and confining stresses and shear strains.

To keep the focus on integrating the dynamical systems model into FEA, the analyses in this paper uses small strain theory in order not to allow the added complexity of large strain deformation, an issue for future work, be a distraction. In addition, the analysis focuses on two-dimensional problems as these are common field problems; the extra complexity that comes from adding a third dimension is again not directly relevant to elucidating the use of the dynamical systems model in a finite element analysis. Likewise, time dependent issues, typically handled using standard methods such as the Crank-Nicholson method, distract from the present focus, and so are not modeled here.

Following standard FEA methodology as used in geotechnical engineering: express the solution for Biot's differential equations for coupled consolidation by finite element discretization in space in terms of nodal forces, displacements and pore pressures. For the case of an incompressible fluid, absent a source or sink for the fluid, the equilibrium and continuity equations are:

$$\partial^T(\bar{\sigma} + u) + F = 0 \tag{2a}$$

$$\mu^T \frac{\partial \delta}{\partial t} - \frac{1}{\gamma f} \Delta^T k \Delta u = 0 \tag{2b}$$

where for the two-dimensional plane strain case commonly encountered in geotechnical engineering field problems:

$$\partial = \begin{pmatrix} \dfrac{\partial}{\partial x} & 0 \\ 0 & \dfrac{\partial}{\partial y} \\ \dfrac{\partial}{\partial y} & \dfrac{\partial}{\partial x} \end{pmatrix}, \quad \Delta = \begin{pmatrix} \dfrac{\partial}{\partial x} \\ \dfrac{\partial}{\partial y} \end{pmatrix}$$

and T applied to any matrix results in its transpose. $\bar{\sigma} = (\bar{\sigma}_x \ \bar{\sigma}_y \ \bar{\sigma}_{xy})$ where $\bar{\sigma}_{xy} = (\bar{\sigma}_x - \bar{\sigma}_y)/2$ are the effective principle stresses in the X (horizontal) and Y (vertical) directions, $F = (F_x \ F_y)$ are the nodal forces (if the body forces are purely self-weight then $F = \gamma_t g$ where γ_t is the total unit mass of the soil and g is a unit vector in the direction of gravity, typically $(0 \ 1)^T$), $\mu = (1 \ 1 \ 0)^T$, the nodal displacements are $\delta = (\delta_x, \delta_y)^T$, t is time, γ_f is the unit weight of the pore fluid, $k = \begin{pmatrix} k_x & k_{xy} \\ k_{yx} & k_x \end{pmatrix}$ is the anisotropic permeability matrix, and u is the pore pressure. See PFEM for details.

Equation 2a is the basic equation used for the undrained triaxial test FEA. Expressing the stress and pore pressure at some step, say i in terms of their existing values, plus the increment of stress during that step, Equation 2a becomes:

$$\partial^T (\bar{\sigma}_i + \Delta\bar{\sigma}_i + u_i + \Delta u_i) + F = 0 \tag{3}$$

where $\Delta\bar{\sigma}_i$ and Δu_i are the increments of effective stress and pore pressure that occur during step i.

For an undrained triaxial compression test, the increment in pore pressure is $\Delta u_i = \Delta\bar{\sigma}_{3i}$.

Now $\Delta\bar{\sigma}_i = \begin{pmatrix} \Delta\bar{\sigma}_{1i} \\ \Delta\bar{\sigma}_{3i} \\ (\Delta\bar{\sigma}_{1i} - \Delta\bar{\sigma}_{3i})/2 \\ \Delta\bar{\sigma}_{3i} \end{pmatrix}$, i.e., for very small increments of shear strain $\Delta\gamma$,

$$\Delta\bar{\sigma}_i = \Delta\gamma_i \begin{pmatrix} d\bar{\sigma}_{1i}/d\gamma \\ d\bar{\sigma}_{3i}/d\gamma \\ (d\bar{\sigma}_{1i}/d\gamma - d\bar{\sigma}_{3i}/d\gamma)/2 \\ d\bar{\sigma}_{3i}/d\gamma \end{pmatrix}$$

Substituting from Equations 1 in terms of the principle stresses σ_1 and σ_3 get:

$$\Delta\bar{\sigma}_i = \Delta\gamma_i \begin{pmatrix} J_q \tan\alpha_q + J_p \tan\alpha_p & -J_q - J_p \\ -J_q \tan\alpha_q + J_p \tan\alpha_p & J_q - J_p \\ J_q \tan\alpha_q & -J_q \\ -J_q \tan\alpha_q + J_p \tan\alpha_p & J_q - J_p \end{pmatrix} \begin{pmatrix} \dfrac{(\bar{\sigma}_1 + \bar{\sigma}_3)}{2} e^{-\lambda_p \gamma_i} \\ \dfrac{(\bar{\sigma}_1 - \bar{\sigma}_3)}{2} e^{-\lambda_q \gamma_i} \end{pmatrix}$$

or:

$$\Delta\bar{\sigma}_i = \Delta\gamma_i (C_{\bar{\sigma}}) \begin{pmatrix} \dfrac{(\bar{\sigma}_1 + \bar{\sigma}_3)}{2} e^{-\lambda_p \gamma_i} \\ \dfrac{(\bar{\sigma}_1 - \bar{\sigma}_3)}{2} e^{-\lambda_q \gamma_i} \end{pmatrix} \tag{4}$$

where:

$$
C_{\bar{\sigma}} = \begin{pmatrix} J_q \tan \alpha_q + J_p \tan \alpha_p & -J_q - J_p \\ -J_q \tan \alpha_q + J_p \tan \alpha_p & J_q - J_p \\ J_q \tan \alpha_q & -J_q \\ -J_q \tan \alpha_q + J_p \tan \alpha_p & J_q - J_p \end{pmatrix}
$$

The increment in pore pressure $\Delta u_i = -\Delta \bar{\sigma}$ or:

$$
\Delta u_i = -\Delta \gamma_i (C_u) \begin{pmatrix} \dfrac{(\bar{\sigma}_1 + \bar{\sigma}_3)}{2} e^{-\lambda_p \gamma_i} \\ \dfrac{(\bar{\sigma}_1 - \bar{\sigma}_3)}{2} e^{-\lambda_q \gamma_i} \end{pmatrix} \tag{5}
$$

where:

$$
C_u = \begin{pmatrix} -J_q \tan \alpha_q + J_p \tan \alpha_p & +J_q - J_p \\ -J_q \tan \alpha_q + J_p \tan \alpha_p & +J_q - J_p \\ 0 & 0 \\ -J_q \tan \alpha_q + J_p \tan \alpha_p & +J_q - J_p \end{pmatrix}
$$

i.e., the total change in stress $\Delta \bar{\sigma}_i + \Delta$ is:

$$
\Delta \sigma_i = \Delta \gamma_i (C) \begin{pmatrix} \dfrac{(\bar{\sigma}_1 + \bar{\sigma}_3)}{2} e^{-\lambda_p \gamma_i} \\ \dfrac{(\bar{\sigma}_1 - \bar{\sigma}_3)}{2} e^{-\lambda_q \gamma_i} \end{pmatrix}
$$

where,

$$
C = C_{\bar{\sigma}} + C_u = \begin{pmatrix} 2J_p \tan \alpha_p & -2J_p \\ 0 & 0 \\ J_q \tan \alpha_q & -J_q \\ 0 & 0 \end{pmatrix}
$$

From equation 3, $\partial^T (\Delta\bar\sigma_i + \Delta u_i) = -F - \partial^T (\bar\sigma_i - u_i)$. Substituting for $\Delta\bar\sigma_i$ and Δu_i we finally obtain our goal–the force equilibrium equation based on the DSSM model:

$$\partial^T \left([C] \left(\begin{array}{c} \dfrac{(\bar\sigma_1 + \bar\sigma_3)}{2} e^{-\lambda_p \gamma_i} \\[2mm] \dfrac{(\bar\sigma_1 - \bar\sigma_3)}{2} e^{-\lambda_q \gamma_i} \end{array} \right) \Delta\gamma_i \right) = -F - \partial^T (\bar\sigma_i + u_i) \tag{6}$$

Standard Galerkin procedures (see for example PFEM) then discretize the terms of this equation, for use in numerical methods. The next section implements this equation in a finite element program, and then uses it to analyze an undrained triaxial test.

EXAMPLE IMPLEMENTATION

Following standard practice, the analysis uses mixed isoparametric quadrilateral elements with 8 nodes (4 corner nodes and 4 mid-span nodes) to calculate displacements, strains, and effective stresses and 4 corner nodes used to determine pore pressure. Displacements are quadratic functions of coordinates, with a different quadratic function for pore pressure. Each stage of the analysis uses standard numerical procedures; treatment of these issues is available in any of the many textbooks on use of finite element methods, such as PFEM. This section describes the implementation of the force equilibrium equation (Equation 6) for the DSSM model, in a FEA program.

The problem analyzed is a simple one, but one which goes to the heart of the matter–that of applying the DSSM constitutive model in a FEA. It is a simple undrained triaxial test where laboratory measurements exist (what the results of the FEA should be). With this basic analysis in hand, applying Equation 6 to more complex problems such as the undrained analysis of field problems becomes merely (as they say in the industry) a question of "time and materials," i.e., given the requisite time, and the materials, there are no unknowns as far as the theory goes, that prevent analyzing the field problem. For the case of drained problems, there is little difference other than additionally applying the Crank-Nicholson routine.

For the analysis, the problem modeled is an undrained triaxial test on a sample 4 inches tall and 2 inches wide. This is an "axisymmetric" problem and so needs analyzing only one-half of the sample cross section. The boundary conditions are that the top and bottom platens are smooth. So, as Figure 8.1 shows, the left-hand boundary (the axis of the cylindrical test specimen) is free to move in the Y direction, but not the X. The bottom is free to move in the X direction, but not the Y. The corner point that is common

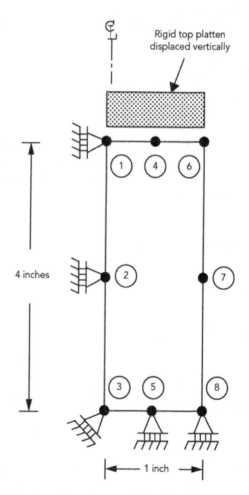

Figure 8.1 Model mesh and boundary conditions (based on Figure 6.49 of PFEM).

to both the axis and the bottom thereby becomes fixed–unable to move in either the X or the Y direction.

Following the earlier thought to focus on the constitutive model itself instead of tangential details, the analyses uses a gingle isoparametric quadrilateral element with 8 nodes (4 corner nodes and 4 mid-span nodes) to calculate displacements, strains, and effective stresses and 4 corner nodes used to determine pore pressure. The element has 4 integration points. Displacements are a quadratic function of coordinates and likewise, with a different quadratic function, for pore pressure. The element has 2 degrees of freedeom at each of its 8 nodes, i.e., 16 degrees of freedom, and, a dimensionality (X,Y) of two. In terms of Equation 6, the stresses and strains are as normally defined for a triaxial test; the value of ∂^T for the axisymmetric 8 node element is the 4 by 16 matrix:

$$
\begin{pmatrix}
\dfrac{\partial N_1}{\partial x} & 0 & \dfrac{\partial N_2}{\partial x} & 0 & \cdots & \dfrac{\partial N_8}{\partial x} & 0 \\[2mm]
0 & \dfrac{\partial N_1}{\partial y} & 0 & \dfrac{\partial N_2}{\partial y} & \cdots & 0 & \dfrac{\partial N_8}{\partial y} \\[2mm]
\dfrac{\partial N_1}{\partial y} & \dfrac{\partial N_1}{\partial x} & \dfrac{\partial N_2}{\partial y} & \dfrac{\partial N_2}{\partial x} & \cdots & \dfrac{\partial N_8}{\partial y} & \dfrac{\partial N_8}{\partial x} \\[2mm]
\dfrac{N_1}{x} & 0 & \dfrac{N_2}{x} & 0 & \cdots & \dfrac{N_8}{x} & 0
\end{pmatrix}
$$

This matrix is discretized as the $[B]$ matrix, whose transpose $[B]^T$ is for this axisymmetric 8 node element, a 16×4 matrix.

Following from this, the stiffness matrix for Equation 6 is:

$$
[B]^T [C] \left(\begin{array}{c} \dfrac{(\bar{\sigma}_1 + \bar{\sigma}_3)}{2} e^{-\lambda_p \gamma_i} \\[3mm] \dfrac{(\bar{\sigma}_1 - \bar{\sigma}_3)}{2} e^{-\lambda_q \gamma_i} \end{array} \right) [B3]
$$

where $[B3]$ is the third row of the B matrix that calculates the shear strain increment when multiplied by the displacement increments at the element nodes, i.e.,

$$
[B3] = \left(\dfrac{\partial N_1}{\partial y} \quad \dfrac{\partial N_1}{\partial x} \quad \dfrac{\partial N_2}{\partial y} \quad \dfrac{\partial N_2}{\partial x} \quad \cdots \quad \dfrac{\partial N_8}{\partial y} \quad \dfrac{\partial N_8}{\partial x} \right)
$$

and

$$
\Delta\gamma_i = [B3](x_1 \ y_1 \ x_2 \ y_2 \ \cdots \ x_8 \ y_8)^T
$$

Inspection of the stiffness matrix shows that it is a 16×16 matrix. Calculating this matrix for each of the four integration points of the element then adding them together results in the total stiffness matrix for the element. If there is more than one element, then assemble the stiffness matrix of each element with contributions from the stiffness matrices of the adjacent elements to obtain the global stiffness matrix. Equate this to the incremental forces on each node, then solve for the incremental X and Y displacements.

The global matrix is non-symmetric and since it can be singular, it is not possible to use the normal Gauss Elimination method to obtain the incremental displacements. Rather, obtain these using the Moore-Penrose pseudo-inverse method–a method that will give the exact solution if the stiffness matrix is non-singular or the least-squares best-fit solution if it is singular, and regardless of whether it is over- or under-determined. Add these displacements to existing nodal shear strains to obtain the total nodal shear strains. Use these to calculate in Equation 6 to calculate the increment in total stress.[5]

The Fortran code that implements the FEA analysis and its associated data file are available for download on the CRC Press web-page for this book. The zip file (triaxial.zip) contains the source code for the analysis.[6] In addition to this file you will need the Moore-Penrose library and three other subroutines, all of which are in the zip file. Compile these three subroutines into the main library and link the triaxial program to the arpack.a, geom.a, main.a and mpinv.a libraries. To run this, download the supporting files from PFEM (available free at: http://inside.mines.edu/~vgriffit/5th_ed/).

As mentioned earlier, example 6.11 of PFEM is the basis of the example in this–so the first thing to do is to get that program working. Whereas the original example 6.11 used the Mohr-Coulomb constitutive model, the modified version uses the DSSM constitutive model. Variables in the program formerly calculated in example 6.11 but now calculated using the DSSM model, keep their old names, but with a "_dssm" tacked on to the end of the variable name. The DSSM program example uses a datafile that calls for 200 increments of vertical displacement that take the specimen to a total vertical strain of 20%. On a very old (circa 2008) and relatively feeble laptop (Lenovo X200, 3 GB RAM, Intel Core2 Duo CPU P8600 @ 2.40GHz) this analysis takes less than a second to run.

The key differences between the DSSM version and example 6.11 that result from replacing 6.11's Mohr-Coulomb failure criterion based model with the DSSM model are that: i) the material matrix is different–it is now a 4×2 matrix that contains the DSSM model's properties for the sample, ii) it recalculates the stiffness matrix at each iteration because the model properties change with strain, iii) the global matrix can be singular and so is solved using the Moore-Penrose pseudo-inverse algorithm discussed earlier and iv) at any given iteration, the standard 4th order Range-Kutta method calculates the new stresses, based on existing stresses and shear strains and the calculated shear strain increment.

---

[5] PFEM uses the routine formtbl to build the global stiffness matrix in the case of non-symmetric stiffness matrices. This global stiffness matrix has only the rows needed for the solution, after considering boundary conditions. Inspection of this matrix shows that the elimination of the leading and trailing zeros, will result in a square matrix, the pseudo-inverse of which, when multiplied by the forces, will return the values the Gaussian elimination method would have, had this matrix been non-singular.

[6] Before you start using these files, make sure you have the latest versions of them. I keep finding errors, mostly minor, and so keep updating them with the latest corrected versions that I have. Should you find an error or scope for improving the code, please be kind enough to let me know so that I can correct my own files.

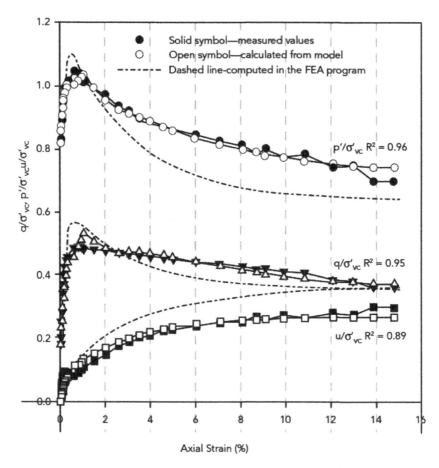

Figure 8.2 Normalized stresses and pore-pressure vs. axial strain; measured, computed directly from the model and from the FEA program (Test CTx-38 from Sheahan, 1991).

Figure 8.2 shows the resulting stress-strain and pore-pressure curves for the element calculated with the program when supplied with the material properties corresponding to Test CTx-38 from Sheahan (1991). This is the same test shown in Figures 2.1 and 3.1 of this book. The specimen tested was a 4″ by 2″ cylinder of Boston Blue Clay-a glacial outwash of illitic USCS-CL clay with approximately 60% greater than clay-size, deposited in a marine environment–one dimensionally (K_o) consolidated to an OCR of 2 and sheared undrained in monotonic compression at a constant axial strain-rate of 0.051%/hour. Also, shown in Figure 8.2 above are the actual values measured by Sheahan, as well as the values computed using the model directly as in Chapter 3, i.e., without the added FEA machinery.

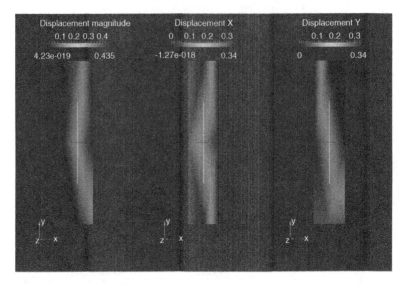

Figure 8.3 Internal displacements–total, x-direction, and y-direction.

As should be the case, the stress-strain and pore-pressure values computed in the FEA program compare reasonably well against the measured values, even post peak strength. This should come as no surprise, as these are the values applied to the specimen and, as for all FEA programs regardless of constitutive model used, are calculated using the constitutive model, which in the case of DSSM, as shown in earlier chapters, matches the test data well. In short, this close match is only a measure of the goodness of the constitutive model and has little if anything to do with FEA per se.

The FEA program can output its results using the Ensight Gold format used by the open source program Paraview, a program that depicts numerical values in the form of easy to interpret figures–see PFEM for details. Figure 8.3 depicts displacements at the end of shear: as a whole, in the X direction, and in the Y direction. The figure clearly depicts the presence of shear planes. However, as mentioned earlier, other than using common sense considerations, strictly speaking, it is not possible to either confirm or deny this result as presently, the required empirical evidence, i.e., test data of actual measurements of internal sample displacements do not exist. Once such data become available, then yes, this limited case does meet the requirements of being falsifiable, i.e., scientific, but until then it still is, strictly speaking, "non-falsifiable." In short, even this limited case of FEA, remains outside the realm of science.

Figure 8.4 shows the internal loads, again in the X direction, and in the Y direction. Here too, the figure depicts the presence of shear planes. Again,

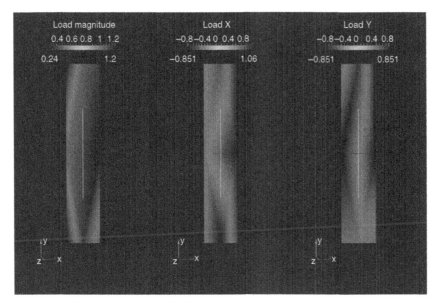

Figure 8.4 Internal loads–total, x-direction, and y-direction.

other than a broad feel for the problem, there is no quantitative basis on which to falsify or confirm this result.

CONCLUSION

This chapter showed how to incorporate DSSM into an FEA program and use it to analyze a well-defined problem. Results matched the empirical data for the problem in terms of overall stress-strain behavior and shed light on how displacements and loads may vary within the sample. The same approach applies to any other undrained problem. For drained cases, additionally, standard numerical methods (ex. Crank-Nicholson) are available to calculate the variation of variables such as stress or pore-pressure with time. For cases where consolidation occurs, the OCR and so the soil properties will vary with time, i.e., the program must recalculate the soil properties each time the OCR changes using the equations for the soil type and stress path, i.e., the corresponding regression equations shown in Figure 3.2 of Chapter 3.

For actual field problems, the variables are many and often unknown. For example, stress paths change continuously, up to and beyond stress reversal. The best way to approach a field problem is to calibrate the model using measurements from the early stages of construction, i.e., the good old "observational method" described in Peck (1969) and pioneered by Terzaghi.

Appendix I

The Poisson process for soil deformation[1]

Often, beginners in a subject are taught key concepts using derivations that though easier to understand (which is why these simplified versions are

---

[1] In 2006, I had already discovered the phenomenological model of Chapter 2. During the review process, Joe Labuz, then editor at the ASCE's Int. Journal of Geotech. and Geoenv. Engg. took me up on my suggestion that an applied mathematician review the paper. The value to me of this anonymous applied mathematician's feedback was worth more than its weight in gold! He said two things that struck me powerfully a) that the model was simple (in the sense that Newton used it when he said, "Nature is simple, and always consonant to itself," i.e., he meant the word as a term of praise!) and b) that yet, while two hundred years ago, a phenomenological model may have sufficed, today, the bar was higher. He asked if I had the physical basis of the model. At that time, I had not even thought of the necessity of driving my phenomenological model down to its physical basis, obvious though it now seems in hindsight. In fact, I did not even know what he meant by "phenomenological" and had to look up the term in a dictionary and introspect deeply on it and its possible connection to my model, as it then existed. Once I understood it, it did not seem likely to me that I ever would be able to figure out the physical basis of the model. Nonetheless, the feedback from Joe Labuz had me primed to think about it, and to be on the lookout for ideas about such a model. A very strict old school professor taught an undergraduate statistics class that I had to take as a "make up" class in the Spring of 2009. He had three exams plus a final. When he taught us the Poisson process, I felt a twinge of recognition. I thought it may have something to do with my phenomenological model, but I was not able to put it together. For the next exam, I tried again to connect the Poisson process to my phenomenological model, but failed again. Then for the final exam, once more I strongly felt this twinge of recognition. I tried a third time ... and failed a third time–I still could not pull it all together. About a month, and many failures later, during Summer school I spent a couple of days and graphed the results of the parameters for Sheahan's data (Figure 3.2 in Chapter 3) incorporating simple friction into the model. I printed out the graph but did not look at it because I was late for a class called "machine learning." So, I simply pulled it from the printer and stuffed it in my bag with my other books and rushed off to class. While in the class though I found that I could not understand what was the teacher was saying, and bored, I surreptitiously pulled out the graph to look at it. Immediately it leaped out at me that the two sets of curves for the material properties (for p' and q) almost

useful) are not quite rigorous. One such concept is that of radioactive decay, taught often as follows: the rate at which particles leave the radioactive material is directly proportional to the number of radioactive particles that make up the material, i.e., $-dR/dt = kR$ where R is the number of radioactive particles in the material, t is time and k, some constant. This, on integrating gives us the well-known exponential decay of radioactivity. Easy, but not rigorous.

Though this concept of radioactivity works in that it results in the correct equation, the derivation is not rigorously based on the underlying physical process that is the root cause of the decay, i.e., the root, physical process that causes radioactive decay is not particles leaving in proportion to their number. The root cause for this behavior is a stochastic process wherein particles leave the radioactive mass at random times.

Processes such as these are "Poisson processes" named after the great French mathematician, **Siméon Denis Poisson** who first discovered them.[2] For soils the Poisson process is this: micro-structural movement of particles

---

overlapped each other. Given that I had no expectation of what I was going to find when I graphed the data, I knew this could not be a coincidence and that the form of the model must be correct. I still did not understand why though and it took me a few more days before it finally all came together and I finally understood how simple friction and the Poisson process both underlay my original phenomenological dynamical systems model. All this happened during the Summer of 2009. I shall always be grateful to Joe Labuz and the anonymous applied mathematician reviewer for their invaluable feedback! For those of you who are curious about it, here is a scan of the printout that so astonished me that day:

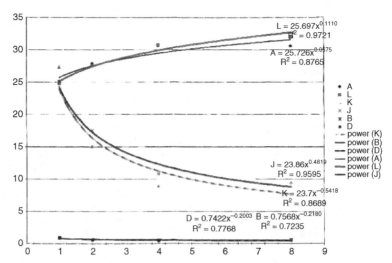

[2]Indeed, Poisson is a true genius–he uncovered an example of nature's sublime simplicity. Think about it–nature puts no restrictions and allows particles to behave freely, randomly ... and out of that, as Poisson discovered, emerges robust order! Exponential decay in this case. The mechanism behind this robust order is so simple

to their final position is a stochastic process in which particles move to their final position at random shear strains. This final structure is the flow-structure for shear to the steady-state or some K structure for shear along some K line. So, while for radioactive decay particles leave at random times, for soil deformation, particles move to their final position at random shear strains. Random time in the case of radioactive decay means that the event of a radioactive particle leaving occurs independent of the time for which the particle was in the radioactive mass. Random shear strain in the case of soil deformation means that the event of the particle moving to the final structure occurs independent of the shear strain for which the particle was not in the final structure.

A little introspection reveals that if particles leave a set of like particles in a random way (as either in the case of radioactive decay or soil deformation), then it follows from this, that the number of particles that leave the set at any given time is proportional to the number of particles there were in the set at that time. In other words, this simpler concept is not wrong, even if it is not rigorously correct, i.e., not the root cause of the manifest/observed phenomena.

So, just as it is not wrong to say for the case of radioactivity that the rate at which particles leave the radioactive material is directly proportional to the number of radioactive particles that make up the material, so also it would not be wrong to make the analogous case for soil deformation–that at any given time or shear strain, particles move to their final position in proportion to the number of particles not in that position at that time or shear strain. In other works, as with the case of radioactive material, so also with soils, i.e., that $-dN/d\gamma = kN$ where N is the number of particles not at the final position, γ is shear strain and k, some constant. This is nothing but the original phenomenological model of Chapter 2. However, it is not the rigorous description though and so it is very important to realize this, and that the root cause is a Poisson process.

---

as to be nearly invisible, requiring a keen, keen nose to smell out (yes, Einstein credited his "nose" for guiding him towards general relativity though of course he did not mean his real nose!). The term "emergent" is very popular these days, especially among non-mathematicians, who use it (accompanied by a lot of "hand waving") to explain everything under the sun, with no appreciation of what it really means or how it really works. Commonly, "consciousness" is thought to be an "emergent" property of the brain. This is obvious, but the question is how exactly? Hand waving and offering "emergence" as an explanation for consciousness simply does not cut it! In the Poisson process, we see a true example of emergence, one that Poisson precisely (mathematically) defined. The Wikipedia definition of "emergence" is a very good one: emergence is a process whereby larger entities, patterns, and regularities arise through interactions among smaller or simpler entities that themselves do not exhibit such properties. We see this clearly in the Poisson process for soils–out of random movements at the microscopic particle level, emerges clear order (exponential decay) of the larger entity. All soil behavior is "emergent" but this time, I use the word with no "hand waving" but instead, in its precise, mathematical sense.

In short, the case of soil deformation is exactly analogous to the case of radioactive decay. Standard books on stochastic processes (see for example Taylor and Karlin, 1998) give the rigorous derivation of radioactive decay for the expected value of the number of radioactive particles that remain at any given time. This appendix reproduces this derivation for the case of radioactive decay, but changes the meaning of the symbols to those pertinent for the case of soil shear, thereby converting the derivation from applying to radioactive decay to now applying to soil deformation. What follows, is this rigorous derivation ... let the statistics begin!

Treat the shear strain for which the particle is not in the flow-structure as a continuous, positive random variable X. Movement into the flow-structure at random shear strains means that the probability of this event occurring at any shear strain is the same, i.e., that $\text{Prob}\{X > \gamma + h \mid X > \gamma\} = \text{Prob}\{X > \gamma\}$, where shear strain h occurs after shear strain γ.

Now,

$$\text{Prob}\{X > \gamma + h\} = \text{Prob}\{X > \gamma + h \mid X > \gamma\}\text{Prob}\{X > \gamma\}$$

$$= \text{Prob}\{X > h\}\text{Prob}\{X > \gamma\}$$

Let $G\{\gamma\} = \text{Prob}\{X > \gamma\}$,

Then $G\{\gamma + h\} = G\{\gamma\}G\{h\}$ and $G'\{\gamma + h\} = G'\{\gamma\}G\{h\} = G\{\gamma\}G'\{h\}$ which means that

$$\frac{G'\{\gamma\}}{G\{\gamma\}} = \frac{G'\{h\}}{G\{h\}} = -\lambda, \text{ i.e., that } G'\{\gamma\} = -\lambda G(\gamma)$$

or that X is exponential, i.e., $\text{Prob}\{X \leq \gamma\} = 1 - G\{\gamma\} = 1 - \exp(-\lambda\gamma)$.

In other words, particles are not in the flow-structure with a probability $\exp(-\lambda\gamma)$, and in the flow-structure with a probability $1 - \exp(-\lambda\gamma)$, where λ is the rate (per unit strain) at which particles move into the flow-structure.

Now, this is a binomial distribution, and hence, for an initial N_0 particles, independent, and identically distributed in X, the probability of N_γ particles not being in the flow-structure at shear strain γ is:

$$\text{Prob}\{N_\gamma = n\} = \frac{N_0!}{n!(N_0 - n)!}\exp(-n\lambda\gamma)[1 - \exp(-\lambda\gamma)]^{N_0-n}$$

where N_γ is the number of particles not in the steady-state position at shear strain γ. So, the expected value and variance of N_γ are $E[N_\gamma] = N_0 \exp(-\lambda\gamma)$ and $\text{Var}[N_\gamma] = N_0 \exp(-\lambda\gamma)[1 - \exp(-\lambda\gamma)]$, and the relative distribution breadth $\text{Var}[N_\gamma]/E[N_\gamma]^2 = [1 - \exp(-\lambda\gamma)]/[N_0 \exp(-\lambda\gamma)]$ which is approximately zero when N_0 is very large.

In short, the expected value of the number of particles not in the steady-state position at any given shear strain γ is $E[N_\gamma] = N_0 \exp(-\lambda\gamma)$.

Appendix 2

Analytical solution to approximated DSSM model

The DSSM model has no analytical solution. However, if the rates λ_q, λ_p at which particles move into the steady-state are equal, then the model does have an analytical solution. The test data from triaxial tests showed that rates λ_q, λ_p were indeed close to each other, and so, at least for the case of triaxial compression, this assumption seems reasonable. This section details the analytical solution for this approximation of the DSSM model; it is based on Joseph and Graham-Eagle (2014).

From the dynamical system equations, it is possible to write $\bar{p}$ and $d\bar{p}/d\gamma$ as functions of q and obtain a single equation in terms of q as

$$\frac{d^2q}{d\gamma^2} + C\frac{dq}{d\gamma} + kq = 0$$

where

$$C = [\lambda_p + J_q \exp(-\lambda_q \gamma) - J_p \tan \alpha_p \exp(-\lambda_q \gamma)]$$

and

$$K = J_q \exp(-\lambda_q \gamma)[(\lambda_p - \lambda_q) + J_p(\tan \alpha_q - \tan \alpha_p) \exp(-\lambda_p \gamma)]$$

The first equation above written in terms of q and the corresponding equation for $\bar{p}$ are each homogeneous, second-order, linear equations with non-autonomous coefficients and have no closed form solution. Given that the rates λ_q, λ_p at which particles move into the steady-state are very close to each other, if we assume that they are equal, it now become possible to analytically solve the equation for q (and likewise if written in terms of $\bar{p}$).

Details of the solution follow below:

Setting $\lambda_q = \lambda_p = \lambda$, the model becomes:

$$\frac{dq}{d\gamma} = J_q \exp(-\lambda\gamma)[\bar{p}\tan\alpha_q - q] \tag{A.1a}$$

$$\frac{d\bar{p}}{d\gamma} = J_p \exp(-\lambda\gamma)[\bar{p}\tan\alpha_p - q] \tag{A.1b}$$

which in matrix notation is:

$$\begin{bmatrix} \dfrac{dq}{d\gamma} \\ \dfrac{d\bar{p}}{d\gamma} \end{bmatrix} = \begin{bmatrix} -J_q & J_q\tan\alpha_q \\ -J_p & J_p\tan\alpha_p \end{bmatrix} \begin{bmatrix} q \\ \bar{p} \end{bmatrix} \exp(-\lambda\gamma) \tag{A.2}$$

or as:

$$X'(\gamma) = \exp(-\lambda\gamma)TX(\gamma) \tag{A.3}$$

where:

$$X(\gamma) = \begin{bmatrix} q \\ \bar{p} \end{bmatrix}, \quad X'(\gamma) = \begin{bmatrix} \dfrac{dq}{d\gamma} \\ \dfrac{d\bar{p}}{d\gamma} \end{bmatrix}, \quad \text{and} \quad T = \begin{bmatrix} -J_q & J_q\tan\alpha_q \\ -J_p & J_p\tan\alpha_p \end{bmatrix}$$

Substituting $X(\gamma) = Z(\exp(-\lambda\gamma))$ in Equation (A.3) $X'(\gamma) = -\lambda\exp(-\lambda\gamma) \cdot Z'(\exp(-\lambda\gamma))$ and substituting these in Equation (A.3) we get:

$$Z'(L) = -\frac{T}{\lambda}Z(L) \tag{A.4}$$

where

$$L = \exp(-\lambda\gamma)$$

or:

$$\begin{bmatrix} Q' \\ \bar{P}' \end{bmatrix} = \begin{bmatrix} \dfrac{J_q}{\lambda} & -\dfrac{J_q\tan\alpha_q}{\lambda} \\ \dfrac{J_p}{\lambda} & -\dfrac{J_p\tan\alpha_p}{\lambda} \end{bmatrix} \begin{bmatrix} Q \\ \bar{P} \end{bmatrix} \tag{A.5}$$

where $q(\lambda) = Q(L)$ and $\bar{p}(\gamma) = \bar{P}(L)$.

Now Equation (A.5) is a homogenous, second-order, linear differential equation. Rewrite it as:

$$Q'' + \frac{1}{\lambda}(J_p \tan\alpha_p - J_q)Q' + \frac{J_q J_p}{\lambda^2}(\tan\alpha_q - \tan\alpha_p)Q = 0 \tag{A.6}$$

This equation has roots:

$$\omega_{1,2} = \frac{-(J_p \tan\alpha_p - J_q) \pm \sqrt{(J_p \tan\alpha_p - J_q)^2 - 4J_q J_p(\tan\alpha_q - \tan\alpha_p)}}{2\lambda}$$

$$\tag{A.7}$$

For the case of real and distinct roots (the case of interest), the standard solution of Equation (A.6) results in:

$$q(\lambda) = C_1 \exp(-\omega_1 \exp(-\lambda\gamma)) + C_2 \exp(-\omega_2 \exp(-\lambda\gamma)) \tag{A.8a}$$

$$\bar{p}(\gamma) = \frac{\lambda}{J_q \tan\alpha_q}\left[C_1\left(\omega_1 + \frac{J_q}{\lambda}\right)\exp(-\omega_1 \exp(-\lambda\gamma)) \right.$$
$$\left. + C_2\left(\omega_2 + \frac{J_q}{\lambda}\right)\exp(\omega_2 \exp(-\lambda\gamma)) \right] \tag{A.8b}$$

References

Aboshi, H. (1973). An experimental investigation on the similitude in the consolidation of a soft clay, including the secondary creep settlement. *Proc. 8th Int. Conf. Soil Mech. Found. Engng*, Moscow 4, No. 3, 88.

Adachi, T., Oka, F., and Mimura, M. (1996). "Modeling aspects associated with time dependent – behavior of soils." In *Session on Measuring and Modeling Time Dependent Soil Behavior, ASCE Convention, Washington*, Geotechnical Special Publication 61, 61–95.

An, J., Tuan, C., Cheeseman, B., and Gazonas, G. (2011). "Simulation of soil behavior under blast loading." *Int. J. Geomech.*, 11(4), 323–334.

Arlitt, Martin F. and Williamson, Carey L. (1997). "Internet web servers: workload characterization and performance implications." *IEEE/ACM Transactions on Networking*, 5(5), 631. doi:10.1109/90.649565.

Arulmoli, K., Muraleetharan, K. K., Hossain, M. M., and Fruth, L. S. (1992). "VELACS verification of liquefaction analyses by centrifuge studies laboratory testing program soil data report." *Rep. Prepared for National Science Foundation*, Earth Technology Corporation, Irvine, CA.

Atkinson, J. H. (1993). *An introduction to the mechanics of soils and foundations*. McGraw-Hill, London, UK.

Billo, J. E. (2007). *Excel for Scientists and Engineers: Numerical Methods-Volume 1*. John Wiley and Sons, N.J.

Bowden, F. P. and Tabor, D. (1950). *The friction and lubrication of solids*. Oxford University Press, Oxford, UK.

Brauer, F. and Bies, D. (2011). *Mathematical models in population biology and epidemiology*. Springer Science + Business Media, NY.

Butterfield, R. (1979). "A natural compression law for soils." *Geotechnique*, 29(4), 469–480.

Cannizzaro, F., Greco, G., Rizzo, S., and Sinagra, E. (1978). "Results of the measurements carried out in order to verify the validity of the Poisson-exponential distribution in radioactive decay events." *The International Journal of Applied Radiation and Isotopes*, 29(11), 649. doi:10.1016/0020–708X(78)90101–1.

Cargill, K. W. (1984). "Prediction of consolidation of very soft soil." *J. Geotech. Eng.*, 110(6), 775–795.

Castro, G. (2016). Personal communications with the author. Emails and telephone convesations; Sept. 24 through Nov. 16, 2016.

Castro, G. (1969). "Liquefaction of sands." Ph.D. Thesis, Harvard University, Cambridge, MA.

Clark, R. W. (1971). *Einstein: the life and times.* World Publishing Company, New York, NY.

Clough, R. W. (1960). "The Finite Element Method in Plane Stress Analysis," *Proc. 2nd ASCE Conf. On Electronic Computation*, Pittsburg, Pa. Sept.

Clough, R. W. (1962). "The Stress Distribution of Norfork Dam," *Institute of Engineering Research, Final Report to the Corps of Engineers*, March 1962, Revised August 1962.

Clough, R. W. and Wilson, E. L. (1999). "Early finite element research at Berkely," *Fifth U.S. National Conference on Computational Mechanics*, Boulder, CO, Aug. 4–6, 1999, USA.

Crawford, M. B. (2010). *Shopcraft as soulcraft: an inquiry into the value of work.* Penguin Books, New York: NY. ISBN: 978-0143117469.

Cundall, P. A., and Stark, O. (1979). A discrete numerical model for granular assemblies. *Geotechnique*, 1, 47–65.

Degago, S. A., Grimstad, G., Jostad, H. P., Nodal, S., and Olsson, M. (2011). "Use and misuse of the isotache concept with respect to creep hypotheses A and B." *Geotechnique*, 61(10), 897–908.

Díaz-Rodríguez, J. A., Martínez-Vasquez, J. J., and Santamarina, J. C. (2009). "Strain-rate effects in Mexico City soil." *J. Geotech. Geoenviron. Eng.*, 135(2), 300–305.

Dowson, D. (1997). *History of Tribology* (2nd ed.). Professional Engineering Publishing (IMechE) ISBN 1-86058-070-X.

Drucker, D. C. and Prager, W. (1952). "Soil mechanics and plastic analysis for limit design." *Quarterly of Applied Mathematics*, 10(2), 157–165.

Feynman, R. (2016). *Feynman on Scientific Method.* http://www.youtube.com/watch?v=EYPapE-3FRw retrieved November 1, 2016.

Finno, J. R., Harris, W. W., and Mooney, M. A. (1996). "Strain localization and undrained steady state of sand." *J. Geotech. Eng.*, 122(6), 462–473.

Fourie, A. B. and Tshabalala, L. (2005). "Initiation of static liquefaction and the role of K_0 consolidation." *Can. Geotech. J.*, 42(3), 892–906.

Fullford, G., Forrester, P., and Jones, A. (1997). *Modelling with differential and difference equations.* Australian Mathematical Society Lecture Series 10, Cambridge University Press, Melbourne, Australia.

Gens, A. (1982). "Stress-strain and strength of a low plasticity clay." Ph.D. Thesis, Imperial College, London.

Gleick, J. (1987). *Chaos: The making of a new science.* Viking Adult, NY.

Gokhale, D. V. (1975). "Maximum Entropy Characterization of Some Distributions." *Statistical Distributions in Scientific Work, Patil, Kotz and Ord. Eds.*, Boston, M.A. Reidel, Vol. 3, 299–304.

Goodman, R. E. (1998). *Karl Terzaghi: The Engineer as Artist*, ASCE, Reston, VA, USA.

Graham, J., Crooks, J. H. A., and Bell, A. L. (1983). "Time effects on the stress-strain behavior of natural soft clays." *Geotechnique*, 33(3), 327–340.

Grimstad, G. (2014). Personal communications with the author. Emails; August 8 through August 16, 2014.

Gunderson, G. H. and Holling, C. S. (2002). *Panarchy: understanding transformations in human and natural systems,* Island Press, Washington, D. C., USA.

Hawlander, B. V., Muhunthan, B., and Imai, G. (2003). "Viscosity effects on one-dimensional consolidation of clays." *Int. J. Geomech.,* ASCE, 3(10), 99–110.

Henkel, D. J. (1956). "The effect of overconsolidation on the behavior of clays during shear." *Geotechnique,* Vol. 6, pp. 139–150.

Henkel, D. J. (1960). "Undrained Shear Strength of Anisotropically Consolidated Clays", *ASCE Speciality Conference on Shear Strength of Cohesive Soils,* University of Colorado, Boulder, Colo., June 13–17, 533–554.

Henkel, D. J. and Wade, N. H. (1966). "Plane Strain Tests on a Saturated Remolded Clay." *J. Soil Mech. and Foundation Div.* SM6, 67–80.

Hicher, P. (1996). "Elastic properties of soils." *J. Geot. and Geonviron. Eng.,* 122(8), 641–648.

Hirschfeld, R. C. (1958). "Factors influencing the constant volume strength of clays." Ph.D. Dissertation, Harvard University, Cambridge, MA.

Hossain, M. and Randolph, M. (2009). "Effect of Strain Rate and Strain Softening on the Penetration Resistance of Spudcan Foundations on clay." *Int. J. Geomech.,* 9(3), 122–132.

Iglesia, G. (2013). Personal communications with the author. Emails; November 11, 2013.

Imai, G., Ohmukai, N., and Tanaka, H. (2004). "An isotaches-type compression model for predicting long term consolidation of KIA clays." In *Proceedings of the Symposium on Geotechnical Aspects of Kansai International Airport,* Kansai, Japan, pp. 49–64.

Imai, G. and Tang, Y. X. (1992). "Constitutive equation of one-dimensional consolidation derived from inter-connected tests." *Soils and Foundations,* 32(2), 83–96.

Jamiolkowski, M., Ladd, C. C., Germain, J. T., and Lancellotta, R. (1985). "New Developments in field and laboratory testing of soils." *Proceedings of the 11th ICSMFE,* San Franscisco, 1, 57–153, Balkema, Rotterdam.

Jaynes, E. T. (1957). "Information theory and statistical mechanics." *Phys. Rev.,* 106, 620–630.

Joseph, P. G. (2009). "A constitutive model of soil based on a dynamical systems approach." *J. Geotech. Geoenviron. Eng.,* ASCE, 135(8), 1155–1158.

Joseph, P. G. (2010). "A dynamical systems based approach to soil shear." *Geotechnique,* LX(10), 807–812.

Joseph, P. G. (2012). "Physical basis and validation of a constitutive model for soil shear derived from micro-structural changes." *Int. J. Geomech.* 13(4), 365–383 (http://dx.doi.org/10.1061/(ASCE)GM.1943–5622.0000209).

Joseph, P. G. (2013a). "Shear modeled from random particle movement." *2nd IACGE International Conference on Geotechnical and Earthquake Engineering,* Chengdu, China. October 25–27, 2013. Proceedings in ASCE Special Pub.

Joseph, P. G. (2013b). "Physical basis of normalizable stress-strain curves for the undrained shear of clays." *66th Canadian Geotechnical Conference,* Montreal, Canada, September 29–October 3, 2013.

Joseph, P. G. (2014a). "Generalized soil deformation model based on dynamical systems theory." *Geotech. Research,* 1(1), 32–42 (http://dx.doi.org/10.1680/geores.14.00004).

Joseph, P. G. (2014b). "Viscosity and Secondary Consolidation in One-Dimensional Loading." *Geotech. Research*, 1(3), 90–98 (http://dx.doi.org/10.1680/gr.14.00008).

Joseph, P. G. and Graham-Eagle, J. (2013). "Strain-Rate Effects in Shear Highlighted by a Dynamical Systems Model." *Int. J. Geomech.*, 14(4), 04014015:1–8 (http://dx.doi.org/10.1061/(ASCE)GM.1943-5622.0000360).

Joseph P. G. and Graham-Eagle, J. (2014). "Analytical solution of a dynamical systems soil model." *14th International Conference of the International Association for Computer Methods and Advances in Geomechanics*, Osaka, Japan, Sept 22–25, in Computer Methods and Recent Advances in Geomechanics, 2015, Editors: F. Oka, A. Murakami, R. Uzuoka, and S. Kimoto, CRC Press, Balkema, Rotterdam, 71.

Karamavruç, A. I. and Clark, N. N. (1996). Application of Deterministic Chaos Theory to Local Instantaneous Temperature, Pressure, and Heat Transfer Coefficients in a Gas Fluidized Bed. *J. Energy Resour. Technol.* 118(3), 214–220, doi:10.1115/1.2793865.

Kim, Y. T. and Leroueil, S. (2001). "Modeling the viscoplastic behavior of clays during consolidation: application to Berthierville clay in both laboratory and field conditions." *Can. Geotech. J.*, 38(3), 484–497.

Lade, P. V. and Yamamuro, J. A. (2011). "Evaluation of static liquefaction potential of silty sand slopes." *Can. Geotech. J.*, 48(2), 247–264.

Ladd, D. C., Foott, R., Ishihara, K., Schlosser, F., and Poulos, H. G. (1977). "Stress-deformation and strength characteristics." *General Report, Proceedings of the 9th International Soil Mechanics and Foundation Engineering Conference*, Vol. 2, 421–494.

LaGatta, D. P. (1970). "Residual strength of clays and clay–shales by rotation shear tests." Ph.D. Thesis, Harvard University, Cambridge, MA.

LaGatta, D. P. (1971). "The effect of rate of displacement on measuring the residual strength of clays." *Contract Report S-71-5 for the U.S. Army Engineer Waterways Experiment Station*, Vicksburg, MI, USA by Harvard University, Cambridge, MA, USA.

Lambe, T. W. (1973). "Predictions in soil engineering." *Géotechnique*, 23(2), 151–202.

Leroueil, S. (2006). "The isotache approach. Where are we 50 years after its development by Professor Šuklje?" *Prof. Šuklje's Memorial Lecture, XIII Danube-European Geotechnical Engineering Conference, Ljubljana, Slovenia*, Ljubljana, Slovenia, 55–88.

Leroueil, S., Kabbaj, M., and Tavenas, F. (1988). "Study of the validity of a RM01_Eqn001.png model in in situ conditions." *Soils and Foundations*, 28(3), 3–25.

Leroueil, S., Kabbaj, M., Tavenas, F., and Bouchard, R. (1986). "Closure to "Stress-strain-strain-rate relation for the compressibility of sensitive natural clays." *Geotechnique*, 36(2), 288–290.

Leroueil, S., Kabbaj, M., Tavenas, F., and Bouchard, R. (1985). "Stress-strain-strain rate relation for the compressibility of sensitive natural clays." *Geotechnique*, 35(2), 159–180.

Li, S., Shirako, H., Sugiyama, M., and Akaishi, M. (2004). "Time effects on one-dimensional consolidation analysis." *Proceedings of the School of Engineering Tokai University*, Vol. 29, 1–8.

Marone, C. (1998a). "Laboratory-derived friction laws and their application to seismic faulting," *Annual Review*, Earth Planetary Science, 643–696.

Marone, C. (1998b). The effect of loading rate on static friction and the rate of fault healing during the earthquake cycle. *Nature*, 391, 69–72.

Mesri, G. (2003). "Primary and secondary compression." *Soil behavior and soft ground construction (eds. Germaine, Sheahan & Whitman)*, ASCE Geotechnical Special Publication 119, 122–166.

Mesri, G. (2009). "Discussion on effects of friction and thickness on long-term consolidation behavior of Osaka Bay Clays." *Soils and Foundations*, 49(5), 823–824.

Mesri, G. and Feng, T. W. (1986). "Stress strain strain rate relation for the compressibility of sensitive natural clays. Discussion." *Géotechnique*, 36(2), 283–287.

Mesri, G. and Castro, A. (1987). "The RM01_Eqn002.png concept and K_0 during secondary compression." *J. Geotech. Eng.*, 113(3), 230–247.

Mesri, G. and Choi, Y. K. (1985). "The uniqueness of the end-of-primary (EOP) void ratio-effective stress relationship." *Proceedings of the 11th ICSMFE*, San Franscisco, Vol. 2, pp. 587–590, Balkema, Rotterdam.

Mesri, G. and Godlewski, P. M. (1977). "Time- and stress-compressibility interrelationship." *Journal of Geotechnical and Geoenvironmental Engineering*, ASCE. 103:417–30.

Mesri, G. and Vardhanabhuti, B. (2009). "Compression of granular materials." *Can. Geotech. J.*, 46(4), 369–392.

Mueller, P. A. and Oppenheimer, D. M. (2014). "The pen is mightier than the keyboard. Advantages of longhand over laptop note taking." Psychological Science, June, Vol. 25, No. 6, pp. 1159–1168. doi:10.1177/0956797614524581.

Nicolis, G. and Prigogine, I. (1977). *Self-Organization in Non-Equilibrium Systems*. Wiley. ISBN 0–471-02401–5.

Niechcial, J. (2002). *A Particle of Clay: The Biography of Alec Skempton, Civil Engineer*. Whittles Publishing. ISBN 1–870325-84–2.

Oka, F. (2005). "Computational modeling of large deformations and the failure of geomaterials." *Theme lecture. Proceedings of the 16th International Conference on Soil Mechanics and Geotechnical Engineering*, Osaka, Vol. 1, 47–95.

Oka, F., Adachi, T. and Okano, Y. (1986). "Two dimensional consolidation analysis using an elasto-viscoplastic constitutive equation," *International Journal for Numerical and Analytical Methods in Geomechanics*, 10(1), 1–16.

Okada, Y., Sassa, K., and Fukuoka, H. (2005). "Undrained shear behaviour of sands subjected to large shear displacement and estimation of excess pore-pressure generation from drained ring shear tests." *Can. Geotech. J.*, 42(3), 787–803.

Peck, R. B. (1969). "Advantages and limitations of the observational method in applied soil mechanics. Ninth Rankine Lecture." *Geotechnique*, 19(2), 171–187.

Plant, J. R. (1956). "Shear strength properties of London Clay." M.Sc. Thesis, Imperial College, London.

Poulos, S. G. (1981). "The steady state of deformation." *J. Geotech. Eng.*, 107(5), 553–562.

Poulos, S. G. (2010). Personal communication to the author dated Oct. 14th, 2010.

Ramaswamy, S. V. R. (2014). Personal communication to the author. Emails between February 2 through February 20, 2014.

Redlich, K. A., Terzaghi, K. v. and Kampe, R. (1929). *Ingenieurgeologie.* Springer-Verlag Wien, Heidelberg, Germany.

Riemer, M. F. and Seed, R. B. (1997). "Factors affecting apparent position of steady-state line," *J. Geotech. Geoenv. Eng.*, 123(3), 281–288.

Roscoe, K. H. and Burland, J. B. (1968). "On the generalized stress-strain behavior of 'wet' clay." *J. Heyman and F. A. Leckie (eds.), Engineering plasticity*, Cambridge University Press, Cambridge, England, pp. 525–609.

Roscoe, K. H. and Schofield, A. N. (1963). "Mechanical behavior of an idealized 'wet' clay." *Proc. European Conf. on Soil Mechanics and Foundation Engineering, Wiesbaden*, Deutsche Gesellschaft fur Erd- und Grundbau e. V., Essen, Vol. 1, 47–54.

Roscoe, K. H., Schofield, A. N., and Wroth, C. P. (1958). "On the yielding of soils", Geotechnique, 8, 22–53.

Schofield, A. N. (2005). *Disturbed soil properties and geotechnical design.* Thomas Telford, London.

Scholz, C. H. (1998). "Earthquake and friction laws–Review Article." *Nature*, 39(1), 37–42.

Sciama, D. W. (1955). "On the formation of galaxies in a steady state universe." Monthly Notices of the Royal Astronomical Society, Vol. 115, pp. 3–14.

Scott, R. F. (1963). *Principles of soil mechanics.* Addison-Wesley, Reading: MA.

Shapiro, S. (2000). "The effects of nonplastic fines on the three-dimensional behavior of sand." M.S. Thesis, Clarkson University, New York.

Sheahan, T. C. (1991). "An experimental study of the time-dependent undrained shear behavior of resedimented clay using automated stress path triaxial equipment." ScD Thesis, Massachusetts Inst. of Tech., Cambridge, MA.

Sheahan, T. C. and Watters, P. J. (1997). "Experimental verification of CRS consolidation theory." *J. Geotech. Geoenviron. Eng.*, 123(5), 430–437.

Skempton, A. (1948). "A study of the geotechnical properties of some post-glacial clays." *Geotechnique*, Vol. 1, pp. 7–22.

Smith, I. M., Griffiths, D. V., and Margetts, L. (2014). *Programming the finite element method.* 5th Edition, John Wiley and Sons, Chichester, Great Britain.

Strogatz, S. H. (1994). *Nonlinear Dynamics and Chaos.* Perseus Publishing Company, N.Y.

Šuklje, L. (1957). "The analysis of the consolidation process by the isotache method." *Proc. 4th Int. Conference on Soil Mechanics and Foundation Engineering*, London, Vol. 1, 200–206.

Tanaka, H., Udaka, K., and Nosaka, T. (2006). "Strain-rate dependency of cohesive soils in consolidation settlement." *Proceedings of the 4th International Conference on Soil Mechanics and Foundation Engineering*, London, Vol. 1, 200–206.

Taylor, H. M. and Karlin, S. (1984). *An introduction to stochastic modeling.* Academic Press, Orlando, FL.

Terzaghi, K. (1925). Erdbaumechanik auf Bodenphysilalischer Grundlage. Franz Deuticke, Liepzig-Vienna.

Terzaghi, K. and Peck, R. B. (1948). *Soil mechanics in engineering practice*. John Wiley and Sons, New York, NY.

Townsend, F. C. (1987). "Symposium on consolidation and disposal of phosphatic and other waste clays." Bartow, FL: Florida Institute of Phosphate Research.

Wang, G. and Sassa, K. (2002). "Post-failure mobility of saturated sands in undrained load-controlled ring shear tests." *Can. Geotech. J.*, 39(4), 821–837.

Watabe, Y., Udaka, K., and Morikawa, Y. (2008). "Strain-rate effects on long-term consolidation of Osaka Bay clay." *Soils and Foundations*, 48(4), 495–509.

Watabe, Y., Udaka, K., Kobayashi, M., Tabata, T., and Emura, T. (2008). "Effects of friction and thickness on long-term consolidation behavior of Osaka Bay clays." *Soils and Foundations*, 48(4), 547–561.

Watabe, Y., Udaka, K., Kobayashi, M., Tabata, T., and Emura, T. (2009). "Closure to effects of friction and thickness on long-term consolidation behavior of Osaka Bay clays." *Soils and Foundations*, 49(5), 824–825.

Watabe, Y., Udaka, K., Nakatni, Y., and Leroueil, S. (2012). "Long-term consolidation behavior interpreted with isotache concept for worldwide clays." *Soils and Foundations*, 52(3), 449–464.

Watabe, Y., Udaka, K., Nakatni, Y., and Leroueil, S. (2013). "Correspondence related to Watabe, Y., Udaka, K., Nakatni, Y., and Leroueil, S. 2012." *Soils and Foundations*, 53(2), 360–362.

Watabe, Y. and Leroueil, S. (2012). "Modeling and implementation of isotache concept for long-term consolidation behavior." *Int. J. Geomech.*, 10.1061/(ASCE)GM.1943–5622.0000270, A4014006.

Willkomm, D., Machiraju, S., Bolot, J., and Wolisz, A. (2009). "Primary user behavior in cellular networks and implications for dynamic spectrum access." *IEEE Communications Magazine*, 47(3), 88, doi:10.1109/MCOM.2009.4804392.

Wroth, C. P. and Bassett, R. H. (1965). "A stress-strain relationship for the shearing behavior of a sand." *Geotechnique*, 15(1), 32–56.

Yamamuro, J. A., Abrantes, A. E., and Lade, P. V. (2011). "Effect of strain rate on the stress-strain behavior of sand." *J. Geotech. Geonviron. Eng.*, 137(12), 1169–1178.

Yamamuro, J. A. and Covert, K. M. (2001). "Monotonic and cyclic liquefaction of very loose sands with high silt content." *J. Geotech. Geonviron. Eng.*, 127(4), 314–324.

Yamamuro, J. A. and Lade, P. V. (1998). "Steady-state concepts and static liquefaction of silty sands." *J. Geotech. Geonviron. Eng.*, 124(9), 868–877.

Yasuhara, K. (1982). "A practical model for secondary compression." *Soils and Foundations*, 48(4), 45–56.

Yin, J. H., Graham, J., Clark, J. L., and Gao, L. (1994). "Modeling unanticipated pore-water pressures in soft clays." *Can. Geotech. J.*, 31(5), 773–778.

Zhu, J., Yin, J., and Luk, S. (1999). "Strain-rate effects on stress-strain strength behavior of soft clay." *Proceedings of the Eleventh Asian Regional Conference on Soil Mechanics and Geotechnical Engineering*, Balkema, Rotterdam, Netherlands, 61–64.

Zhu, J. and Yin, J. (2000). "Strain-rate dependent stress-strain behavior of over consolidated Hong Kong marine clay." *Can. Geotech. J.*, 37(6), 1272–1282.

Subject index

Note: Page numbers in **bold** indicate figures